bras

P9-AEZ-694

Polymers in Sensors

ACS SYMPOSIUM SERIES **690**

Polymers in Sensors

Theory and Practice

Naim Akmal, EDITOR
Union Carbide Technical Center

Arthur M. Usmani, EDITOR
Usmani Development Company

Developed from a symposium sponsored by the
Division of Industrial and Engineering Chemistry
at the 212th National Meeting
of the American Chemical Society,
Orlando, Florida,
August 25–29, 1996

American Chemical Society, Washington, DC

Library of Congress Cataloging-in-Publication Data

Polymers in Sensors : theory and practice / Naim Akmal, editor, Arthur M. Usmani, editor

p. cm.—(ACS symposium series, ISSN 0097–6156; 690)

"Developed from a symposium sponsored by the Division of Industrial and Engineering Chemistry at the 212th National Meeting of the American Chemical Society, Orlando, Florida, August 25–29, 1996."

Includes bibliographical references and indexes.

ISBN 0–8412–3550–3

1. Chemical detectors—Congresses. 2. Biosensors—Congresses. 3. Polymers—Congresses.

I. Akmal, Naim, 1962–   . II. Usmani, Arthur M, 1940–   . III. American Chemical Society. Division of Industrial and Engineering Chemistry. IV. American Chemical Society. Meeting (212th : 1996: Orlando, Fla.) V. Series.

TP159.C46P66   1997
681'.2—dc21
                                                                98–13989
                                                                    CIP

This book is printed on acid-free, recycled paper.

Copyright © 1998 American Chemical Society

Distributed by Oxford University Press

All Rights Reserved. Reprographic copying beyond that permitted by Sections 107 or 108 of the U.S. Copyright Act is allowed for internal use only, provided that a per-chapter fee of $20.00 plus $0.25 per page is paid to the Copyright Clearance Center, Inc., 222 Rosewood Drive, Danvers, MA 01923, USA. Republication or reproduction for sale of pages in this book is permitted only under license from ACS. Direct these and other permissions requests to ACS Copyright Office, Publications Division, 1155 16th Street, N.W., Washington, DC 20036.

The citation of trade names and/or names of manufacturers in this publication is not to be construed as an endorsement or as approval by ACS of the commercial products or services referenced herein; nor should the mere reference herein to any drawing, specification, chemical process, or other data be regarded as a license or as a conveyance of any right or permission to the holder, reader, or any other person or corporation, to manufacture, reproduce, use, or sell any patented invention or copyrighted work that may in any way be related thereto. Registered names, trademarks, etc., used in this publication, even without specific indication thereof, are not to be considered unprotected by law.

PRINTED IN THE UNITED STATES OF AMERICA

TP159
C46 P66
1998
CHEM

Advisory Board

ACS Symposium Series

Mary E. Castellion
ChemEdit Company

Arthur B. Ellis
University of Wisconsin at Madison

Jeffrey S. Gaffney
Argonne National Laboratory

Gunda I. Georg
University of Kansas

Lawrence P. Klemann
Nabisco Foods Group

Richard N. Loeppky
University of Missouri

Cynthia A. Maryanoff
R. W. Johnson Pharmaceutical
    Research Institute

Roger A. Minear
University of Illinois
    at Urbana–Champaign

Omkaram Nalamasu
AT&T Bell Laboratories

Kinam Park
Purdue University

Katherine R. Porter
Duke University

Douglas A. Smith
The DAS Group, Inc.

Martin R. Tant
Eastman Chemical Co.

Michael D. Taylor
Parke-Davis Pharmaceutical
    Research

Leroy B. Townsend
University of Michigan

William C. Walker
DuPont Company

Foreword

THE ACS SYMPOSIUM SERIES was first published in 1974 to provide a mechanism for publishing symposia quickly in book form. The purpose of the series is to publish timely, comprehensive books developed from ACS sponsored symposia based on current scientific research. Occasionally, books are developed from symposia sponsored by other organizations when the topic is of keen interest to the chemistry audience.

Before agreeing to publish a book, the proposed table of contents is reviewed for appropriate and comprehensive coverage and for interest to the audience. Some papers may be excluded in order to better focus the book; others may be added to provide comprehensiveness. When appropriate, overview or introductory chapters are added. Drafts of chapters are peer-reviewed prior to final acceptance or rejection, and manuscripts are prepared in camera-ready format.

As a rule, only original research papers and original review papers are included in the volumes. Verbatim reproductions of previously published papers are not accepted.

ACS BOOKS DEPARTMENT

Contents

ION-SELECTIVE ELECTRODES AND POLYMER-BASED SENSORS AND BIOSENSORS

FIBER OPTIC SENSORS

INDEXES

Preface

THE FIELD OF SENSING TECHNOLOGY covers a vast area of expertise and application in various arenas. The sensor is a logical element in the informations-acquisition chain. Sensors provide information about our physical, chemical, and biological environments. The rapid growth in technology and its application has created a major market for various kinds of sensing devices to maintain the high quality of the final product and simultaneously to increase the yield.

There is no doubt that chemical sensors and biosensors are fast-moving, critical technologies for industrial and biomedical marks. Sensors find wide applications in medicine (for example, blood chemistry determinations and immunological and microbiological testing), food, agriculture, and environmental and industrial monitoring. The value of the industrial gas sensor industry is projected to more than double by 1998, with semiconductor sensors and electrochemical sensors leading the way. The total market for worldwide chemical sensors was $700 million in 1994. This market is growing at an estimated rate of 9.0% per annum. Also, efforts to reduce the risk of cross-contamination and physician liability risks are opening up opportunities for the manufacturers of disposable biomedical sensors.

The symposium was organized to address the latest developments in sensing technology and its application in various industries. This volume presents the missing link in chemical and biosensing technology and improvements performed in recent years. The chapters cover a wide arena of sensors used in chemical and other related industries.

The chapters in this book have been divided into four sections: diagnostics and biosensors, gas sensors and their applications, ISE and polymer-based sensors and biosensors, and fiber optic sensors. Chapter 1 provides an overview of the diagnostic section that has not been covered in any other chapter. Chapters 2–7 deal with biosensors and their biomedical applications. Chapters 8–17 provide excellent information pertaining to various gas-sensing technologies for process industries. Chapters 18–21 provide an overview of the applications of polymers in sensing technology. A separate section, chapters 22–23, has been allotted to fiber optic sensors.

This book should be useful to chemists, biochemists, chemical engineers, process engineers, polymer scientists, and materials scientists. It provides a good source of information to anyone interested in learning about gas sensors—their chemistry and their limitations. Scientists interested in doing research and development work in diagnostics and biosensors can take advantage of the book's broad coverage of sensing technology.

Acknowledgments

We sincerely thank all those who contributed to the successful publication of this volume. We express our appreciation to Teledyne Electronic Technologies and Usmani Development Company (UDC) for their support. Also we acknowledge the support provided by the staff of ACS Books. The gracious support of our families is most warmly acknowledged.

NAIM AKMAL
Union Carbide Technical Center
3200 Kanawha Turnpike
South Charleston, WV 25303

ARTHUR M. USMANI
Usmani Development Company
7318 Normandy Way
Indianapolis, IN 46278

DIAGNOSTICS AND BIOSENSORS

Chapter 1

An Overview of Medical Polymers and Diagnostic Reagents

Naim Akmal[1] and A. M. Usmani[2]

[1]Union Carbide Technical Center, 3200 Kanawha Turnpike,
South Charleston, WV 25303
[2]Usmani Development Company, 7318 Normandy Way, Indianapolis, IN 46278

This work describes the principles and biochemical reactions involved in diagnostic reagents, dry chemistry construction and recent advances in biosensors. Among all the methods and techniques available to date for the accurate and fast detection of blood sugar and cholesterol, dry chemistry is still the number one choice. Blood glucose test strips are made up of suitable enzymes along with indicators such as 3,3'-5,5'-tetramethylbenzidine in the form of a thin film and layered over a plastic film. Importance and selection of polymer binders which play a major role in dry chemistry has been discussed along with the thermal analysis data and its role for various diagnostic enzymes.

Biosensors using enzyme, GOD for monitoring glucose and the use of long chain polymers to wire the enzyme to the electrode, in order to have fast electron movement is also discussed.

Purified enzymes are invariably used in medical diagnostic reagents and in the measurement of analytes in urine, plasma, serum or whole blood. There has been a steady growth of dry chemistry during the past three decades. It has surpassed wet clinical analysis in the number of tests performed in hospitals, laboratories and homes because of its ease, reliability and accuracy.

Enzymes are specific catalysts that can be derived from plant and animal tissues; however fermentation continues to be the most popular method. Enzymes are extensively used in diagnostics, immunodiagnostics, and biosensors. They measure or amplify signals of many specific metabolites. Purified enzymes are expensive and

2

their use for a large number of analytes can be expensive. This is the main reason for the increasing use of reusable immobilized enzymes in clinical analyses.

Wet chemistry methods for analysis of body analytes, e.g., blood glucose or cholesterol requires equipment and trained analysts. Millions of people with diabetes check their blood glucose levels and are able to obtain results in a matter of a few minutes. However, science has not yet invented an insulin delivery system that can respond to the body's senses. Injected insulin does not automatically adjust and, therefore, the dose required to mimic the body's response must be adjusted daily or even hourly depending on the diet and physical activity. Self-monitoring of blood glucose levels is essential for diabetics. This has become possible since the last 30 years due to the advent of dry chemistry (1-8). Accurate monitoring of blood glucose level by an expectant woman will enable her to have normal pregnancy and give birth to a healthy child. Athletes with diabetes can self-test their blood glucose levels to avoid significant health problems. Dry chemistries are useful not only to diabetes, but also to patients with other medical problems. They are also used in animal diagnosis, food evaluation, fermentation, agriculture, and environmental and industrial monitoring (1).

Biochemical Reactions

Biochemical reactions for assaying cholesterol and glucose are shown below:

Cholesterol

$$\text{Cholesterol esters} + H_2O \xrightarrow{\text{Cholesterol esterase}} \text{Cholesterol} + \text{Fatty acids}$$

$$\text{Cholesterol} + O_2 \xrightarrow{\text{Cholesterol oxidase}} \text{Cholest-4-en-3-one} + H_2O_2$$

$$H_2O_2 + \text{chromogen} \xrightarrow{\text{Peroxidase}} \text{Dye} + H_2O$$

Notes: (1)End-point followed by dye formation.
(2)Amount of oxygen consumed can be measured amperometrically by an oxygen-sensing electrode.
(3)The H_2O_2 produced by cholesterol oxidase requires phenol to produce dye. A popular alternative step is to substitute p-hydroxybenzenesulfonate for phenol in the reaction with pyridine nucleotide. The subsequent reaction is as follows:

$$H_2O_2 + \text{ethanol} \xrightarrow{\text{catalase}} \text{acetaldehyde} + 2H_2O$$

$$\text{Acetaldehyde} + NAD(P)^+ \xrightarrow{\text{aldehyde dehydrogenase}} \text{acetate} + H^+ + NAD(P)H$$

(4)Free cholesterol determined, if cholesterol esterase omitted.

Glucose

(1) $$\text{Glucose} + O_2 + H_2O \xrightarrow{\text{Glucose oxidase}} \text{Gluconic acid} + H_2O_2$$
$$H_2O_2 + \text{Chromogen} \xrightarrow{\text{Peroxidase}} \text{Dye} + H_2O_2$$
Note: Formation of dye.

(2) $$\text{Glucose} + O_2 + H_2O \xrightarrow{\text{Glucose oxidase}} \text{Gluconic acid} + H_2O_2$$
Note: The rate of oxygen depletion is measured with an oxygen electrode. Two additional steps prevent the formation of oxygen from H_2O_2 and are as follows.

$$H_2O_2 + \text{ethanol} \xrightarrow{\;catalase\;} \text{acetaldehyde} + 2H_2O$$

$$H_2O_2 + 2H^+ + 2I^- \xrightarrow{\;molybydate\;} I_2 + 2H_2O$$

Immobilized Enzymes in Diagnostic Reagents

During the past 30 years immobilized enzymes have received considerable attention in comparison to soluble enzymes for measurement of analytes. Use of immobilized enzymes offer the advantages of accuracy, greater stability and convenience.

There is considerable literature available describing the application of immobilized enzymes and analysis. Some of them have been described in references (9-14). However, only a few of these methods have been commercialized. Furthermore, immobilized enzymes have not yet achieved their full potential in clinical chemistry. Dry chemistry has outpaced all the combined clinical chemistry methods. This is, described in detail in the next section. Biosensors are also gaining momentum and their prospects are now promising due to advances in redox polymers.

Advances in polymers have resulted in many successful implants, therapeutic devices and diagnostic assays (15). Polymers can be manufactured in a wide range of composition and their properties can be regulated by composition variations and modifications. Furthermore, they can be configured ranging from simple to complex shapes. Polymer-enhanced dry chemistries and biosensors are gaining importance and will continue to grow in the future. Polymers are used to enhance speed, sensitivity and versatility of both biosensors and dry chemistries in diagnostics to measure vital analytes. Today's diagnostic medicine is placing demands on technology for new materials and application techniques. Polymers are thus finding ever increasing use in diagnostic medical reagents (7). A comprehensive discussion of dry chemistry is presented in the following paragraphs.

Dry Chemistry

History. Dry or solid phase can trace its origins back to the ancient Greeks. Some 2000 years ago, copper sulfate was an important ingredient in tanning and preservation of leather. At that time dishonest traders were adulterating valuable copper sulfate with iron salts. Pliny has described a method for soaking reeds of papyrus in a plant gall infusion or a solution of gallic acid that would turn black in the presence of iron. This is the first recorded use of the dry chemistry system. In 1830, filter paper impregnated with silver carbonate was used to detect uric acid qualitatively.

Self-testing, e.g., measuring ones own weight and body temperature is likewise an old concept. Elliot Joslin advocated some 80 years ago that diabetics should frequently monitor their glucose level by testing their urine with Benedict's qualitative test (16). When insulin became available in the early 1920s, self testing of urine became necessary. In the mid-1940s, Compton and Treneer compounded a dry tablet consisting of sodium hydroxide, citric acid, sodium carbonate and cupric sulfate (17). This tablet when added to a small urine sample and boiled resulted in the reduction of blue cupric sulfate to yellow or orange color, if glucose was present. Glucose urine strips, impregnated and based on enzymatic reactions using glucose oxidase, peroxidase and an indicator were introduced around 1956 (18). Dry chemistry blood glucose test strips coated with a semipermeable membrane to which whole blood could be applied and wiped-off were introduced in 1964. Currently, low-cost, lightweight plastic-housed reflectance meters with advanced data management capabilities are available and being extensively used.

Impregnated dry reagents have a coarse texture with high porosity and an uneven large pore size, resulting in a non-uniform color development in the reacted strip. In the early 1970s a coating-film type dry reagent was developed by applying an enzymatic coating onto a plastic support. This gave a smooth, fine texture and therefore a uniform color development (19). Nonwipe dry chemistries wherein, the devices handle the excess blood by absorption or capillary action have appeared in the marketplace during the last 5 years.

Discrete multi-layered coatings, developed by the photographic industry were adapted in the late 1970s to coat dry reagent chemistry formats for clinical testing (20, 21). Each zone of the multi-layered coating provides a unique environment for sequential chemical and physical reactions. These devices consist of a spreading layer, separation membrane, reagent, and reflective zone, coated onto the base support. The spreading layer wick's the sample and applies it uniformly to the next layer. The separation layer can hold back certain components, e.g., red blood cells allowing only the described metabolites to pass through to the reagent zone that contains all the necessary reagent components. Other layers maybe incorporated for filtering, or eliminating the interfering substrates.

Background Enzymes are essential in dry chemistries. In a typical glucose measuring dry reagent, glucose oxidase (GOD) and peroxidase (POD) enzymes, along with a suitable indicator, e.g., 3,3',5,5'-tetramethylbenzidine (TMB), are dissolved and/or dispersed in a latex or water soluble polymer. This coating is then applied to a lightly pigmented plastic film and dried to obtain thin dry film. The coated plastic can be cut to a suitable size, 0.5 cm x 0.5 cm and mounted on a plastic handle. The user applies a drop of blood on the dry reagent pad and allows it to react for about 60 s or less. The blood can be wiped off manually by a swab or by the device itself. The developed color is then read by a meter or visually compared with the pre-designated printed color blocks to determine the precise glucose level in the blood. Thus, dry chemistries are highly user friendly.

Dry chemistry test kits are available in thin strips that are usually disposable. They may be either film-coated or impregnated. The most basic diagnostic strip consists of a paper or plastic base, polymeric binder, and reacting chemistry components consisting of enzymes, surfactants, buffers and indicators. Diagnostic coatings or impregnation must incorporate all reagents necessary for the reaction. The coating can either be single or multi-layer in design. A list of analytes, enzymes, drugs and electrolytes assayed by dry chemistry diagnostic test kits is given in Table I (22,23).

Table I. Dry Chemistry Diagnostic Test Kits

Substrate	Enzymes	Drugs	Electrolytes
Glucose		Phenobarbitone	
Urea		Phenytoin Theophyline	Sodium ion
Urate		Carbamazepine	Chloride
Cholesterol (total)	Alkaline Phosphate (ALP)		CO_2
Triglyceride	Lactate dehyrogenase (LDH)		
Amylase (total)	Creatine kinase MB isoenzyme (CK-MB)		
Bilirubin (total)	Lipase		
Ammonium ion Creatinine Calcium Hemoglobin HDL cholesterol Magnesium (II) Phosphate (inorganic) Albumin Protein in cerebrospinal fluid			

Dry chemistry systems are being widely used in physician's offices, hospital laboratories and by millions of patients worldwide. They are used for routine urine analysis, blood chemistry determinations, and immunological and microbiological testing. The main advantage of dry chemistry technology is that it eliminates the need for reagent preparation and many other manual steps common to liquid reagent systems, resulting in greater consistency and reliability of test results. Furthermore, dry chemistry has a longer shelf stability and therefore helps to reduce wastage of reagents. Each test unit contains all the reagents and reactants necessary to perform the assays.

Dry chemistry tests are used for the assay of metabolites by concentration or by activity in a biological matrix. In general, reactive components are usually present in excess, except for the analyte being determined. This is done to make sure that the reactions will be completed quickly. Other enzymes or reagents are used to drive the reactions in the desired direction (24). Glucose and cholesterol are the most commonly used measured analytes. The biochemical reactions in dry chemistries for many analytes including glucose and cholesterol have been described earlier.

Components of the Dry Chemistry System The basic components of typical dry reagents that utilize reflectance measurement are a base support material, a reflective layer, and a reagent layer which can either be single-or multi-layered. The functions of the various building blocks are now detailed. The base layer serves as a building base for the dry reagent and is usually a thin rigid thermoplastic film. The reflective layer, usually white pigment filled plastic film, coating, foam, or paper reflect not absorbed by the chemistry to the detector. The reagent layer contains the integrated reagents for a specific chemistry. Typical materials are paper matrix, fiber matrix, coating-film as well as combinations thereof.

An example of a single-layer coating reagent that effectively excludes red blood cells (RBC) is shown in Figure 1. Here, an emulsion based coating containing all the reagents for specific chemistry, is coated onto a lightly filled thermoplastic film and dried. For glucose measurement, the coating should contain GOD, POD, and TMB. It may also contain buffer for pH adjustment, minor amounts of ether-alcohol type organic coalescing agent, and traces of a hindered phenol type antioxidant to serve as a color signal ranging compound. Whole blood is applied and allowed to react for 60 s or less; excess blood is wiped off, and the developed color is read visually or by a meter. Paper impregnated type dry chemistry is shown in Figure 2.

The schematic of a typical multi-layered coating dry reagent, for example blood urea nitrogen is shown in Figure 3. It consists of a spreading and reflective layer. A sample containing BUN is spread uniformly on this layer. The first reagent layer is a porous coating-film containing the enzyme urease and a buffer (pH 8.0). Urease reduces BUN to NH_3. A semi-permeable membrane coating allows NH_3 to permeate while excluding OH$^-$ from the second reagent layer. The second reagent layer is composed of a porous coating-film containing a pH indicator wherein, the indicator color develops when NH_3 reaches the semi-permeable coating-film. Typically, such dry reagents are slides (2.8 x 2.4 cm) with an application area of 0.8 cm^2 and the spreading layer is about 100 mm thin.

Application of Diagnostic Technology in Monitoring Diabetes Frequent measurement of blood glucose to manage diabetes is one of the most important applications of diagnostic reagents. Researchers are making new inroads in understanding diabetes whereas, medicinal chemists are exploring new drugs to treat the disease and its complications. At the same time, biotechnologists are discovering better ways to manage the disease.

It has been estimated that there are about 15 million diabetics in the USA but only half of them have been diagnosed with the disease. More than 1.5 million diabetics are treated with injected insulin and the rest with a weight loss program, diet, and oral antidiabetic drugs, e.g., sulfonylureas (Tolbutamide, Tolzamide, Chloropropamide, Glipizide, Glyburide).

The current US market for drugs to control blood glucose is about one billion dollars, equally divided between insulin and antidiabetic drugs (25). Insulin will grow by about 10% annually, whereas, the antidiabetic drug market will shrink by about 3%. The blood glucose monitoring market is about $750 million in the US, and is expected to grow at a rate of 10% annually.

8

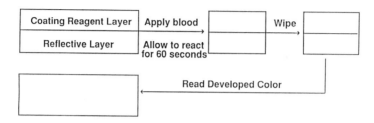

Figure 1. Coating-film type dry chemistry.

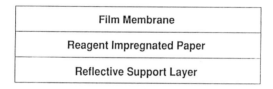

Figure 2. Paper impregnated and overcoated type dry chemistry.

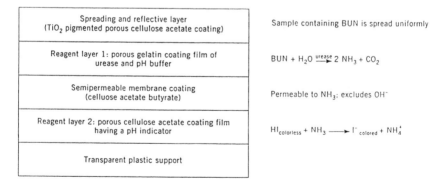

Figure 3. Example of multi-layered coating dry chemistry for blood urea nitrogen.

Diagnostic Polymers and Coatings   In most dry chemistries, polymers account for more than 95% of the strips. Polymers are therefore, important and must be selected carefully. Usmani and coworkers have linked polymer chemistry to biochemistry, and that has lead to a better understanding of and improvement in dry chemistries. General considerations of polymers in dry chemistries are described in the following paragraphs.

The polymer binder is necessary to incorporate the system's chemistry components in the form of either a coating or impregnation. The reagent matrix must be carefully selected to mitigate or eliminate non-uniformity in reagents' concentrations due to improper mixing, settling, or non-uniform coating thickness. Therefore, aqueous based emulsion polymers and water soluble polymers are being extensively used. Table II provides a list of commonly used matrix binders (26,27).

Table II. Common Matrix Binders

Emulsion Polymers	**Water Soluble Polymers**
Acrylic	Polyvinyl alcohol
Polyvinyl acetate: homo and co-polymers	Polyvinyl pyrrolidone Highly hydroxylated acrylic
Styrene acrylics	Polyvinylethylene glycol acrylate
Polyvinyl propionate: homo and co-polymers	Polyacrylamide Hydroxyethyl cellulose
Ethylene vinyl acetate	Other hydrophilic cellulosics
Lightly crosslinkable acrylics	Various co-polymers
Polyurethanes	

Polymers must be carefully screened and selected to avoid interference with the chemistry. The properties of polymers, e.g., composition, solubility, viscosity, solid content, surfactants, residual initiators, film-forming temperature and particle size should be carefully considered (28). In general, the polymer should be a good film-former with good adhesion to the support substrate. Furthermore, it should have no or minimal tack for handling purposes, during manufacturing of the strip. The coated matrix or impregnation must have the desired pore size and porosity to allow penetration of the analyte being measured as well as have the desired gloss, swelling characteristics, and surface energetics. Swelling of the polymer binder due to the absorption of the liquid sample may or may not be advantageous depending on the system. Emulsion polymers have a distinct advantage over soluble polymers, due to their high molecular weight, superior mechanical properties, and potential for adsorbing enzymes.

Polymeric binders used in multi-layered coatings include various emulsion polymers, gelatin, polyacrylamide, polysaccharides, e.g., agarose, water soluble polymers, e.g., polyvinyl pyrrolidone, polyvinyl alcohol, copolymers of vinyl pyrrolidone and acrylamide, and hydrophilic cellulose derivatives, e.g., hydroxyethyl cellulose and methyl cellulose.

Water-borne Coatings This type is by far the most important coating-film in use toady because enzymes are water soluble. During the late-80's we researched a tough coating-film in which the whole blood was allowed to flow over the coating film. This is shown in Figure 4. RBC must roll over and not stick to the coating film. Almost all types of emulsions and water soluble polymers were investigated. A styrene-acrylate emulsion was found to be most suitable. A coating containing 51 g styrene-acrylate emulsion (50% solids), 10 g linear alkylbenzene sulfonate (15 wt% in water), 0.1 g GOD, 0.23 g POD, 0.74 g TMB, 10.0 g hexanol, and 13.2 g 1-methoxy-2-propanol gave a good dose response as seen in Figure 5. This coating film gave a linear dose response, in transmission mode, up to 1500 mg/dL of glucose. The coating was slightly tacky and was found to be highly residence and time dependent.

To mitigate or eliminate tack, ultra fine mica was found to be effective. This made possible the rolling and unrolling of coated plastic for automated strip construction. In devices wherein blood residence time is regulated, the residence time dependent coating will be of no problem. Coating-film dependency creates a serious problem where blood residence time cannot be regulated by the strip. To make the coating-film residence time independent, continuous leaching of color from films during exposure time is essential. Enzymes and the indicator can leach rapidly near the blood/coating-film interface and develop color that can be drained off. The color formed on the coating-film is then independent of the exposure time. It was found that this gave a coating-film with minimum time dependency. A low molecular weight polyvinyl pyrrolidone (e.g., PVP K-15 from GAF or PVP K-12 from BASF) gave coatings with no residence time dependency. Such a coating consists of 204 g styrene-acrylate emulsion (50% solids), 40 g linear alkybenzene sulfonate (15% solution in water), 0.41 g GOD (193 U/mg), 1.02 g POD (162 U/mg), 2.96 g TMB, 7.4 g PVP K-12, 32 g micromica C-4000, 20 g hexanol and 6 g Igepal CO-530. The nonionic surfactant serves as a surface modifier to eliminate RBC retention.

Excellent correlation was found when results at 660 nm and 749 nm were correlated with a reference hexokinase glucose method. The dose response was excellent up to 300 mg/dL glucose. In general, water-borne coatings do not lend themselves to ranging by antioxidants.

Non-aqueous Coatings For the past 25 years, dry reagent coatings have exclusively been water-borne due to the belief that enzymes function only in a water medium. Recently, non-aqueous enzymatic coatings for dry chemistries have been researched, developed and refined by Usmani et al (24, 29). The red blood cells will

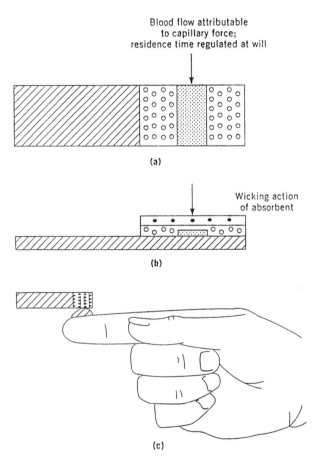

Figure 4. Touch and drain dry chemistry construction. (a) Dry coated surface, (b) cross-section of dry coated surface, adhesive and cover piece, (c) contact with blood drop results in blood filing the cavity. After desired reaction time, blood is drained off by touching end of the cavity with absorbent material.

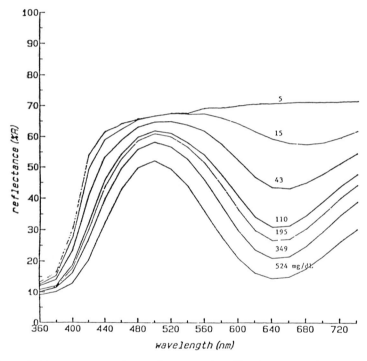

Figure 5. Dose response of water-borne coating-film for 120 s residence time.

not adhere to such chemistries and they gave a quick end-point. These coatings also gave superior thermostability. Furthermore, these coatings could be easily ranged by antioxidants whereas, water-borne coatings are very difficult to range. These researchers synthesized non-aqueous hydroxyalted acrylic diagnostic polymers with good hydrophilicity and hydro-gel character. The enzymes, e.g., GOD and POD are insoluble in organic solvents but become extremely rigid and can be dispersed with ease. Dispersions of less than 1 mm were made using an Attritor mill or a ball mill. To prepare non-aqueous coatings, polymer solution, TMB, mica, surface modifiers and solvents were added to the enzyme dispersion and mixed slightly on a ball mill. The ranging compound can be post-added. The wt% composition of a typical non-aqueous coating useful for a low range blood glucose measurement is 33.29 hydroxylethyl methacrylate/butyl methacrylate/dimethylaminoethyl methacrylate (65/33/2) polymer (40% solid), 2.38 TMB, 1.17 GOD, 2.68 POD, 3.28 sodium dodecyl benzene sulfonate, 26.53 xylene, 26.53 1-methoxypropanol, and 1.96 cosmetic grade C-4000 ultra-fine mica. Many surfactants and surface modifiers were investigated that eliminated RBC retention. Antioxidants that function as ranging compounds in these non-aqueous systems include 3-amino-9-(aminopropyl)-carbazole dihydrochloride, butylated hydroxy toluene, and a combination of BHT/propyl gallate. These ranging compounds are effective in ranging compound to TMB indicator molar ratio of 1:2.5 to 1:20.

The long-term stability of the non-aqueous coatings-films under elevated temperature and moderate humidity has been reported to be better than aqueous coatings. Furthermore, color resolution and sensitivity of reacted nonaqueous coating-films were excellent as demonstrated by the curve fitting.

Molded Dry Chemistry In general, most enzymes are very fragile and sensitive to pH, solvent, and elevated temperatures. Typical properties of select diagnostic enzymes are given in Table III. The catalytic activity of most enzymes is reduced dramatically as the temperature is increased. Common enzymes used in diagnostics, e.g., GOD and POD are almost completely deactivated around 65°C in solid form or as a water solution. Despite wide and continued use of such enzymes in diagnostics for more than 40 years, limited or no thermal analysis work on these biopolymers have been reported until recently (30). Recent differential scanning calorimetric (DSC) analysis results indicating glass transition temperature (T_g), melting temperature (T_m), and decomposition temperatures (T_d) are shown in Table IV. Below T_g, the enzymes are in a glassy state and should be thermally stable. Around T_g, onset of the rubbery state begins, and the enzyme becomes prone to thermal instability. When the enzymes melt around T_m, all the tertiary structures are destroyed, thus making the enzyme completely inactive. The presence of chemicals can considerably influence enzyme stability.

Table III. Typical Properties of Select Diagnostic Enzymes

	Cholesterol Oxidase (CO)	Cholesterol Esterase (CE)	Glucose Oxidase (GOD)	Peroxidase (POD)
Source	Streptomyces	Pseudomonas	Aspergillus	Horseradish
EC	1.1.3.6	1.1.13	1.1.3.4	1.11.1.7
Molecular Weight	34,000	300,000	153,000	40,000
Isoelectric Point	5.1 ± 0.1 & 5.4 ± 0.1	5.95 ± 0.05	4.2 ± 0.1	
Michaelis Constant	4.3×10^{-5} M	2.3×10^{-5} M	3.3×10^{-2} M	
Inhibitor	Hg^{++}, Ag^{+}	Hg^{++}, Ag^{+}	Hg^{+}, Ag^{+}, Cu^{++}	CN^{-}, S^{--}
Optimum pH	6.5 - 7.0	7.0 - 9.0	5.0	6.0 - 6.5
Optimum Temp.	45 - 50°C	40°C	30 - 40°C	45°C
pH Stability	5.0 - 10.0 (25°C, 20h)	5.0 - 9.0 (25°C, 20h)	4.0 - 6.0 (40°C, 1h)	5.0 - 10.0 (25°C, 20h)

Table IV DSC Analysis of Select Diagnostic Enzymes

Enzyme	Source	T_g(°C)	T_m(°C)	T_d(°C)
Cholesterol Oxidase	Nocardia	50	98	210
Cholesterol Oxidase	Streptomyces	51	102	250
Cholesterol Esterase	Pseudomonas	43	88	162
Glucose Oxidase	Aspergillus	50	105	220
Peroxidase	Horseradish	50	100	225

The redox centers (FAD/FADH2) in the GOD enzyme can conduct electrons and are catalytically relevant. To keep or sustain the enzyme's activity, the redox centers must remain intact. The bulk of the enzyme, which is polymeric in composition, is an insulator; altering it, will not reduce the enzyme's catalytic activity. Recently, an enzymatic compound containing GOD, POD, TMB, a linear alkylbenzene sulfonate, and polyhydroxy ethyl methacrylate was molded between 105 to 150°C. This gave a good response to glucose (34). Molding at 200°C resulted in enzyme deactivation. A mechanism has been proposed whereby the enzymes are protected by the tight PHEMA coils. These researchers further suggest that molding of strips using RIM may lead to useful chemistries, including biosensors in the future.

Biosensors

Biosensors and DNA probes are a few new trends in diagnostic reagents. Work is currently underway on the near infrared method (NIR) which will be a reagentless system as well as non-invasive. Prospects for the reagentless NIR method for glucose and other analytes is uncertain at this time.

In the early 1960s, Clark investigated a promising approach to glucose monitoring in the form of an enzyme electrode that used oxidation of glucose by the enzyme GOD (32). This approach has been incorporated and used in a few clinical analyzers for blood glucose determination. There are about three detection approaches in the glucose enzyme electrode (Figure 6). Oxygen consumption is measured in the first method. A reference, nonenzymatic electrode is required, however, to provide an amperometric signal. The second approach detects H_2O_2 but requires an applied potential of about 650 mV and an inside permselective membrane. In the third generation biosensor, advantage is taken of the fact that the enzymatic reaction happen in two steps. The GOD enzyme is reduced by glucose and then the reduced enzyme is oxidized by an electron acceptor, i.e., a mediator most specifically a redox polymer. Direct electron transfer between GOD and the electrode occurs extremely slowly; therefore an electron acceptor mediator is required to make the process rapid and effective (33).

During the past several years, much work has been done on exploration and development of redox polymers that can rapidly and efficiently shuttle electrons. Several research groups have "wired" the enzyme to the electrode with a long chain polymer having a dense array of electron relays. The polymer penetrates and binds the enzyme, and is also bound to the electrode. Heller and his group have done extensive work on the Os-containing polymers. They have made a large number of such polymers and evaluated their electrochemical characteristics (34). Their most stable and reproducible redox polymer is a poly(4-vinyl pyridine) to which $Os(bpy)_2Cl_2$ has been attached to $1/6^{th}$ of the pendant pyridine groups. The resultant redox polymer is water insoluble. To make it water soluble and biologic compatible, Heller et al have

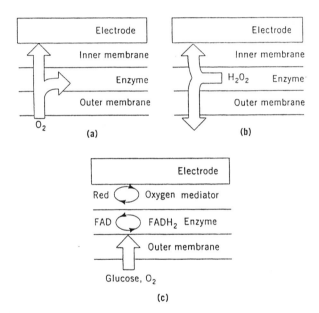

Figure 6. Detection methods for glucose enzyme electrode based on (a) oxygen, (b) peroxide, and (c) a mediator.

partially quaternized the remaining pyridine pendants with 2-bromoethyl amine. This polymer is now water soluble and the newly introduced amine groups can react with a water soluble epoxy, e.g., polyethylene glycol diglycidyl ether and GOD to produce a crosslinked biosensor coating-film. Such coating-films produce high current densities and a linear response to glucose up to 600 mg/dL. The synthesis and application of osmium polymers have been refined by Usmani and coworkers (35). They have also successfully made osmium monomers that should also shuttle electrons much like the polymer version (39, 40). A schematic depiction of osmium polymer and polymer/GOD hydro-gel films useful in biosensors is shown in Figure 7 (36).

Okamoto and his group used a flexible polymer chain in contrast, to put relays (37, 38). Their polymers provide communication between GOD's redox centers and the electrode. No mediation occurred when ferrocene was attached to a non-silicone backbone. Their ferrocene-modified siloxane polymers are stable and non-diffusing. Therefore, biosensors based on these redox polymers gave a good response and are superior in stability. Recently, commercial electrochemical micro-biosensors, e.g., Exactech (Medisense) and silicone based 6+ system (i-Stat) have appeared in the marketplace. These new technologies will certainly impact rapid blood chemistry determinations by the turn of this century. A typically important example is that of blood glucose determination on very small blood volumes (<5 µL) obtained by a finger-stick. This is possible, since detection instruments can be designed more compactly than opto-electronic systems. Certainly, there is a market for small, disposable electrochemical tests in the emergency room, surgical and critical care units as well as homes.

In a biosensor we seek to compete effectively with GOD's cofactor, namely, oxygen for the transfer of redox equivalents from the active site to the electrode surface. A good choice for the artificial electron relay depends on a molecule's ability to reach the reduced $FADH_2$ active site, undergo fast electron transfer, and then transport this to the electrodes as rapidly as possible. Surridge and coworkers have done electron-transport rate studies in an enzyme electrode for glucose using Interdigitated Array Electrodes (39). In addition, Surridge has proposed the following mechanism in osmium polymer/GOD biosensor films.

$$GOD(FAD) + G \xrightarrow{k1} GOD(FAD).G \xrightarrow{k2} GOD(FADH_2) + GL$$

$$GOD(FADH_2) + 2Os\,(III) \xrightarrow{k3} GOD(FAD) + 2Os\,(II) + 2H^+$$

$$Os(II)_1 + Os(III)_2 \xrightarrow{ke} Os(III)_1 + Os(II)_2$$

$$Os(II)_2 \xrightarrow{Fast} Os(III)_2 + e^-$$

Wherein;

Os (II)	reduced form
Os(III)	oxidized form
G	glucose
GL	glucolactone

Schuhmann has recently suggested that the next generation of amperometric enzyme electrodes will be based on immobilization techniques that are compatible with microelectronics mass production processes and will be easy to be miniaturized (40). Integration of enzymes and mediators simultaneously should improve the electron-transfer pathway from the active site of the enzyme to the electrode. In this work, Schuhmann deposited functionalized conducting monomers on electrode surfaces aiming for covalent attachment or entrapment of sensor components. Conductive polymers, e.g., polypyrrole, polyanaline and polythiophene are formed at the anode by electrochemical polymerization. For integration of bioselective compounds and/or redox polymers into conductive polymers, functionalization of conductive polymer films is essential. In Figure 8 schematic representation of an amperometric biosensor with the enzyme covalently bound to a functionalized conductive polymer, e.g., β-amino(polypyrrole), Poly[N-(4-aminophenyl)-2,2'-dithienyl]pyrrole is shown. Entrapment of ferrocene-modified GOD within the polypyrrole is shown in Figure 9. Finally, schematic representation of an amperometirc biosensor with a pyrrole-modified enzyme copolymerized with pyrrole is presented in Figure 10.

There is a pressing need for an implantable glucose sensor for optimal control of blood glucose concentration in diabetics. A biosesnor providing continuous readings of blood glucose will be most useful at the onset of hyper- or hypoglycemia, enabling a patient to take corrective measures. Furthermore, incorporating such a biosensor into a close-loop system with a microprocessor and an insulin infusion pump could provide automatic regulation of the patient's blood glucose. Morff et al used two novel technologies in the fabrication of a miniature electroenzyme glucose sensor for implantation in the subcutaneous tissues of humans with diabetes (41). They developed an electrodeposition technique to electrically attract GOD and albumin onto the surface of the working electrode. The resultant enzyme/albumin layer was crosslinked by butraldehyde. They also developed a biocompatible polyethylene glycol/polyurethane copolymer to serve as the outer membrane of the sensor to provide differential permeability of oxygen relative to glucose to avoid the oxygen deficit encountered in physiologic tissues.

Concluding Remarks

We have provided a comprehensive review of the science and technology of medical diagnostic reagents including wet chemistry, dry chemistry, and biosensors. Despite advances in biosensors, dry chemistry will dominate the reagent market due to economic reasons. Needless to say, the practice of medicine will be very difficult, if not impossible, without medical diagnostic reagents. Biochemists and biotechnologists have kept pace with the needs of the medical industry. Medical diagnostic technology is moving fast and we can expect better, faster, easier to use novel reagents. At any rate, these reagents have definitley improved the quality of life of the afflicted.

20

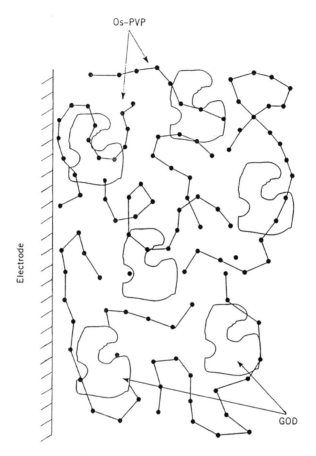

Figure 7. Structure of Os-conducting redox polymer and depiction of polymer/GOD hydrogel film on a biosensor.

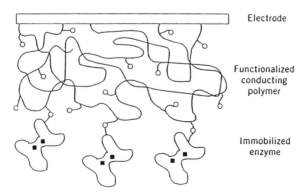

Figure 8. Schematic representation of enzyme covalently bound to a functionalized conductive polymer.

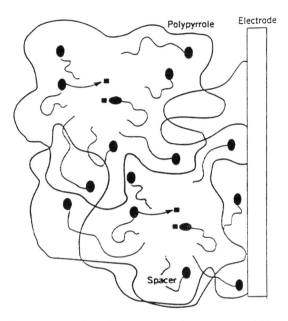

Figure 9. Entrapment of mediator-modified enzymes within a conductive polymer film.

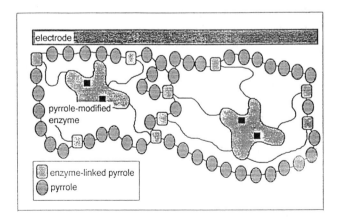

Figure 10. Schematical representation of an amperometric biosensor with a pyrrole-modified enzyme coploymerized with pyrrole.

22

References

1. Walter, B. *Anal. Chem.* **1983**, *55*, 449A.
2. Free, A. H.; Free, H. M. *Lab. Med.* **1984**, *15*, 1595.
3. Mayer, T. K.; Kubasik, N. K. *Lab. Mgmt.* **1986**, 43.
4. Jackson, J. A.; Conard, M. E. *Am. Clin. Products Rev.* **1987**, *6*, 10.
5. Azhar, A. F; Burke, A. D.; DuBois, J. E.; Usmani, A. M. *Polymer Mater. Sci. Eng.* **1988**, *59*, 1539.
6. Diebold, E.; Rapkin M.; Usmani, A. M. *Chem Tech.* **1991**, *21*, 462.
7. Skarstedt, M. T.; Usmani, A. M. *Polymer News* **1989**, *14*, 38.
8. Campbell, R. S.; Price, C. P. *J. Intern. Fed. Clin. Chem.* **1991**, *3*, 204.
9. Carr, P. W.; Bowers, L.D. *Immobilization of Enzymes in Analytical and Clinical Chemistry*; Wiley: 1980, New York.
10. Guilbault,G. G *Handbook of Enzymatic Methods of Analysis*; Dekker: New York, 1976.
11. Free, A. H. *Ann. Clin. Lab Sci.* **1987**, *7*, 479.
12. Ngo, T. T. *Int. J. Biochem.* **1980**, *11*, 459.
13. Bowers, L. D. *Trends Anal. Chem.* **1982**, *1*, 191.
14. Mottola, H. A. *Anal. Chim. Acta.* **1983**, *145*, 27.
15. Azhar, A. F.; Usmani, A. M. *Handbook of Polymer Degradation*; Dekker: New York, 1992.
16. Joslin, E. P.; Root, H. P.; White, P.; Marble, A. *The Treatment of Diabetes*, 7th *Edition*: Philadelphia, 1940.
17. Compton, W. A.; Treneer J. M. *US Patent 2,387244* **1945**.
18. Free, A. H.; Adams, E. C.; Kercher, M. L. *Clin. Chem.* **1957**, *3*, 163.
19. Rey, H. G.; Rieckmann, P.; Wielinger, H.; Rittersdorf, W. *US Patent 3,630975* **1971**, assigned to Boehringer Mannheim.
20. Shirey, T. L. *Clin. Biochem.* **1983**, *16*, 147.
21. Przybylowicz, E. P.; Millikan, A. G. *US Patent 3,992157* **1976**, assigned to Eastman Kodak.
22. Spiegel, H. E. *Kirk-Othmer Encyclopedia of Chemical Technology*, *3rd Edition*, Wiley: New York, 1985.
23. Usmani, A. M.; *Diagnostic Biosensor Polymers*; ACS Symposium Series 556; ACS: Washington, D.C., 1994: Chapter 1, pp 2.
24. Tietz, N. W. *Fundamental of Clinical Chemistry*; W. B. Saunders: Philadelphia, 1976.
25. Stinson, S. C. *C&EN* **1991**, *September 30*, 635.
26. Usmani, A. M. *Biotechnology Symposium*, **1992**.
27. Bruschi, B. J. *US Patent 4,006403* **1978**, assigned to Eastman Kodak.
28. Scheler, W. *Makromol. Chem. Sym.* **1987**, *12*, 1.
29. Azhar, A. F.; Usmani, A. M.; Burke, A.D.; DuBois-Bousamra, J.; Diebold, E.R.; Rapkin, M. C.; Skarstedt,M. T. *US Patent 5,260195* **1993**, assigned to Boehringer Mannheim.
30. Kennamer, J. E.; Usmani, A. M. *J. App. Polym. Sci.* **1991**, *42*, 3073.
31. Kennamer, J. E.; Burke, A.D.; Usmani, A. M. *Biotechnology and Bioactive Polymers*; Plenum: New York, 1992.

32. Clark, L. C.; Lyons, C. *Ann. NY Acad. Sci.* **1962**, *102*, 29.
33. Reach, G.; Wilson, G. S. *Anal. Chem.* **1992**, *64*, 381A.
34. Gregg, B. A.; Heller, A. *J. Phy. Chem.* **1991**, *95*, 5970.
35. Usmani, A. M.; Deng, D. *Unpublished work* **1992**, Boehringer Mannheim.
36. Usmani, A. M.; Surridge, N. A; Diebold, E. R. *Unpublished work* **1992**, Boehringer Mannheim.
37. Boguslavsky, L.; Hale, P.; Skotheim, T.; Karan, H.; Lee, H.; Okamoto,Y. *Polym. Mater Sci. Eng.* **1991**, *64*, 322.
38. Karan, H. I.; Lan, H. L.; Okamoto, Y. *Diagnostic Biosensor Polymers*; ACS Symposium Series 556; ACS: Washington, D.C., 1994; Chapter 14, pp 169.
39. Surridge, N. A.; Diebold,E. R.; Chang, J.; Neudeck, G. W. *Diagnostic Biosensor Polymers*; ACS Symposium Series 556; ACS: Washington, D.C., 1994; Chapter 5, pp 47.
40. Schehmann, W.; *Diagnostic Biosensor Polymers*; ACS Symposium Series 556; ACS: Washington, D.C., 1994; Chapter, 9, pp 110.
41. Johnson, K. W.; Allen, D. J.; Mastrototaro, J. J.; Morff, R. J.; Nevin, R. S. *Diagnostic Biosensor Polymers*; ACS Symposium Series 556; ACS: Washington, D.C., 1994; Chapter, 7, pp 84.

Chapter 2

Biocompatible Microdialysis Hollow-Fiber Probes for Long-Term In Vivo Glucose Monitoring

Kazuhiko Ishihara[1], Nobuo Nakabayashi[1], Michiharu Sakakida[2], Kenro Nishida[2], and Motoaki Shichiri[2]

[1]Institute for Medical and Dental Engineering, Tokyo Medical and Dental University, Tokyo 101, Japan
[2]Department of Metabolic Medicine, Kumamoto University School of Medicine, Kumamoto 860, Japan

A microdialysis system is powerful tool for monitoring biosubstances in living organisms. This system needs a semipermeable hollow fiber probe to collect the biosubstances and to eliminate proteins and cells which interfere with the monitoring. However, the performance of the hollow fiber probe decreased drastically when the probe is inserted into the tissue. Therefore, the surface biocompatibility of the hollow fiber probe must be improved for long-term continuous monitoring. We have synthesized a water-soluble graft polymer composed of a cellulose and polymer having a phospholipid polar group, poly[2-methacryloyl-oxyethyl phosphorylcholine (MPC)](MGC) as a coating material on the cellulose hemodialysis membrane. The MGC could be coated on the hollow fiber probe and the performance of the probe did not decrease even after the coating. By recording rapid changes in the glucose concentrations from 100 to 200 mg/dL, the time to reach 90 % of the maximum value was within 7 min. To determine the glucose concentration in subcutaneous tissues, the hollow fiber probe modified with the MGC was applied to human volunteers. The glucose recovery through the original hollow fiber probe was not significantly different during 4 days of continuous monitoring, however, then significantly decreased at day 7 compared with its original value. On the other hand, using the MGC-modified hollow fiber probe, the glucose recovery was not significantly different during 7 days of monitoring. The surface observation of the hollow fiber probe after insertion indicated that both protein adsorption and cell adhesion were suppressed by the MGC coating and this is one of the reasons for improving the stability of the monitoring system *in vivo*.

In 1982, Shichiri *et al.* succeeded in miniaturizing the glucose monitoring system with a needle-type glucose sensor used *in vivo* studies(*1*). A wearable artificial endocrine pancreas was then developed. This system consisted of a needle-type glucose sensor, microcomputer system, insulin and glucagon infusion pump system and a battery. The entire system was packaged as a small unit. It was first

©1998 American Chemical Society

demonstrated that physiological glycemic regulation for a long period could be obtained with this system in ambulatory diabetic patients. However, for long-term clinical application of a wearable artificial endocrine pancreas, there are many problems to be solved as shown in Table 1, especially, for a glucose sensing system. A major obstacle in extending the life-time of the subcutaneous tissue glucose monitoring was the failure to develop a stable and reliable glucose monitoring system. It is considered that the major reason for a decay in the sensing system would possibly be protein adsorption on the membrane surface when implanted into the subcutaneous tissue. To prolong the life of the glucose monitoring system, the biocompatible membrane becomes crucial(2).

To extend the life-time of the glucose monitoring system, a newly designed biocompatible polymer membrane was constructed using 2-methacryloyloxyethyl phosphorylcholine(MPC)(3). It was found that the MPC copolymers with n-butyl methacrylate(BMA) showed improved biocompatibility(4-7). The poly(MPC-co-BMA) suppressed blood cell adhesion and activation when the polymer contacted human platelet-rich plasma and whole blood even in the absence of an anticoagulant. The amount of plasma proteins adsorbed on the MPC polymer from human plasma was drastically reduced with an increase in the MPC composition in the polymer. These results clearly indicated that the MPC moieties in the polymer play an important role in improved blood compatibility. The needle-type implantable glucose sensor coated with poly(MPC-co-BMA) functioned well for more than 14 days even when it was implanted in subcutaneous tissue(8). To create a stable and reliable glucose monitoring system, a miniaturized extracorporeal glucose monitoring system based on the microdialysis sampling method was then considered(9). The microdialysis sampling method is a promising way for monitoring low-molecular weight biosubstances in living organism. This system needs a semipermeable hollow fiber probe to separate the target biosubstances from proteins. However, the performance of the hollow fiber probe drastically decreased when the probe is inserted in the tissue. Therefore, the surface biocompatibility of the hollow fiber probe must be improved for long-term continuous monitoring with the system.

The MPC undergo polymerization in water and be grafted to a cellulose membrane surface even in a heterogeneous system(10,11). Therefore, the surface of cellulose membranes for hemodialysis was modified with poly(MPC). The biocompatibility test results indicated that the introduction of poly(MPC) chains on the membrane surface was effective in preventing platelet adhesion, complement activation, and plasma protein adsorption.

We considered that a water-soluble polymer having both an affinity to the cellulose base substrate and excellent biocompatibility could improve the surface biocompatibility of the cellulose membrane using a convenient and safe technique, such as coating from an aqueous solution(12). Therefore, we synthesized the graft polymer composed of a cellulose backbone and introduced the poly(MPC) chains. This communication describes the improved performance on a cellulose hollow fiber probe for a microdialysis system using MGC and the clinical application in a novel glucose monitoring system which can be continuously used *in vivo* for long time.

Experimental

Materials. Methacrylate with a phosphorylcholine group, MPC, was synthesized using a previously reported method(13). The water soluble graft polymer consisting of cellulose and poly(MPC) was synthesized(*12*). The chemical structure of the poly(MPC)-grafted cellulose (MGC) is shown in Figure 1. The MPC mole fraction in MGC was 0.34.

Surface modification of cellulose flat membrane with MGC. The cellulose membrane (Cuprophan, thickness, 20 µm, Enka A.G., Wappertal-Barmen, Germany) for hemodialysis was immersed into an aqueous solution containing 0.5 wt% MGC for 3 min, dried at room temperature, and dried in vacuo. The coating of the membrane with the MGC was confirmed by X-ray photoelectron spectroscopy(XPS, ESCA-750, Shimadzu, Kyoto, Japan). The MPC mole fraction on the membrane surface was 0.25 determined from XPS data. The amount of MGC coated on the cellulose membrane was 48 µg/cm^2 determined by phosphorus analysis.

Permeation of glucose was evaluated by the following experimental method using a U-shaped glass cell at 37 °C. The original cellulose membrane or that coated with the MGC saturated with water was interposed between two parts of the glass cell. The initial concentrations of glucose were 200 mg/dL. The concentration of glucose permeated through the membrane was determined using a clinical test kit.

Preparation of cellulose follow fiber probe coated with the MGC for microdialysis. The microdialysis sampling system was purchased from Eicom, Kyoto, Japan. The microdialysis hollow fiber probe is made of cellulose, Cuprophane, and the length of the sampling part was 15 mm with a molecular weight cut-off of 5.0 x 10^4. The surface of the hollow fiber probe was coated with MGC using the solution dipping method with a 0.5 wt% aqueous solution of MGC.

The principle of this monitoring system is dialysis. An illustration of the system is shown in Figure 2. The hollow fiber probe was implanted in the subcutaneous tissue and perfused with physiological saline solution at a rate of 120 µL/h. The saline solution comes to the hollow fiber, while the dialysate comes directly to the glucose electrode. The glucose concentration in the dialysate is continuously determined with an extracorporeal needle-type glucose sensor.

Measurement of subcutaneous glucose concentration using microdialysis system. After the reliability of this monitoring system was guaranteed for at least 7 days, the continuous subcutaneous tissue glucose monitoring in healthy and diabetic volunteers was carried out. The cellulose hollow fiber probe was connected by a tube to a microdialysis unit, in which a flow-cell with a needle-type glucose sensor for extracorporeal sensing, a micro-roller pump, saline reservoir and waste bag are incorporated. Sensor currents are then transmitted to the glucose monitoring system. Subcutaneous tissue glucose concentrations are digitally displayed on a continuous basis using this monitoring system. In this system, hyperglycemia or hypoglycemia is notified by an alarm sound. In this experiment, the cellulose hollow fiber probe was inserted into the subcutaneous tissue of the abdomen, and the perfusion rate was selected at 120 µL/h. A picture of the system is shown in Figure 3.

Results and Discussion

Fundamental properties of cellulose membrane coated with MGC. As shown in Figure 4, when MGC was immobilized on the surface of the cellulose membrane from an aqueous solution after drying, the cellulose backbone in the MGC could act

Table I. Long-term clinical application of glucose sensor problems awaiting solution

1. Maintenance of excellent sensor characteristics
   Dependence on oxygen tension   - Application of electron mediator
   Dependence on temperature - Application of thermister
2. Noise elimination
   Development of noise filter
3. Longevity
   Selection of biomaterials and membrane design
   Development of extracorporealblood glucose monitoring system
   Development of internal calibration system
4. Interface between the living organism and sensor
   Safety for immobilized enzymes
   Immunological response to biomaterials
5. Light and small system
6. Non-invasive measurement
7. Low cost

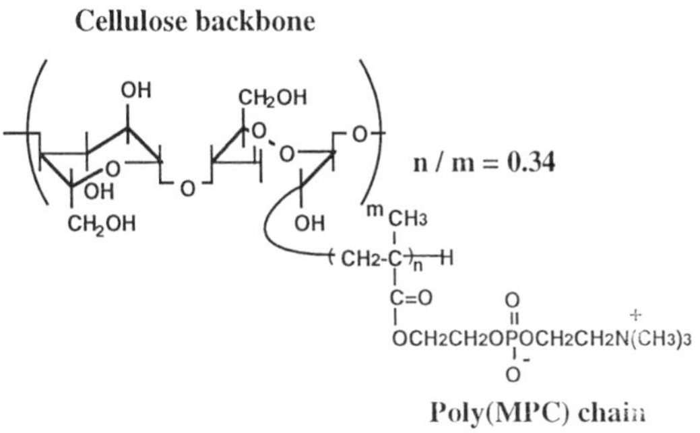

Figure 1. Chemical structure of water-soluble graft polymer composed of cellulose and poly(MPC)(MGC). The MGC can be used as a biocompatible coating of cellulose substrate.

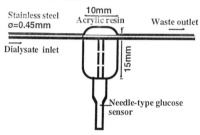

(A) Hollow fiber probe

Polyimide-coated fused silica ø=0.15mm

Stainless steel ø=0.45mm

Hollow fiber ø=0.22mm

Outlet cannula

15mm

Stainless steel ø=0.65mm

Acrylic resin

Inlet cannula

(B) Flow cell unit with sensor

Stainless steel ø=0.45mm

10mm Acrylic resin

Waste outlet

Dialysate inlet

15mm

Needle-type glucose sensor

Figure 2. Schematic diagram of a cellulose hollow fiber probe and a sensor flow cell in the miniaturized extracorporeal glucose monitoring system based on the microdialysis sampling method.

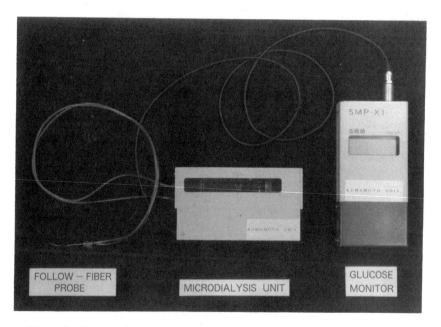

FOLLOW – FIBER PROBE

MICRODIALYSIS UNIT

GLUCOSE MONITOR

Figure 3. Picture of extracorporeal glucose monitoring system developed at Kumamoto University, Japan.

as a fixation site for the poly(MPC) chains on the surface because of the strong affinity due to hydrogen bonding between the base cellulose and the cellulose backbone. That is, the cellulose backbone could be selectively oriented to the base cellulose membrane and the poly(MPC) chains could spread out when blood contacts the membrane. Since the treatment involves only coating and drying, it provides a simple process to improve blood compatibility of the cellulose membrane. Moreover, the synthetic reaction of the MGC can proceed homogeneously, and the yield of the MGC was high.

Surface characterization of the membrane was performed with XPS. Though the original cellulose membrane had XPS signals attributed to carbon and oxygen atoms, the signals attributed to the nitrogen atom(404 eV) and phosphorus atom(134 eV) were observed on the surface of the membrane coated with MGC(12). The ratio of nitrogen atom and phosphorus atom was almost 1.0. This suggests that the MPC units were located at the surface of the membrane. Elution of the MGC was evaluated after the cellulose membranes were immersed in water at 37 °C for 3 h and 100 °C for 30 min. The values were low compared to the total amount of MGC on the membrane, the value was less than 10 % of the initial values. The MPC mole fraction at the membrane surface in the dry state(0.25) was slightly smaller than that in the bulk polymer(0.34). However, after immersion in water, the MPC mole fraction at the surface increased to 0.39.

The tensile strength of the original cellulose membrane in the wet state was 1.4 ± 0.2 kgf/mm^2 and that coated with the MGC was 1.3 ± 0.1 kgf/mm^2. There was no significant difference between them($p > .01$). The permeation amount of glucose through the cellulose membrane coated with MGC linearly increased with time. From these results, we concluded that the modification of the cellulose membrane with MGC would not have any adverse effect of membrane performance.

Property of cellulose hollow fiber probe covered with MGC. In the preliminary experiments, the glucose recovery in the dialysate was inversely proportional to the perfusion rate. The absolute glucose recovery in the dialysate, expressed as mg glucose per hour, was zero at zero flow rate, reached a broad maximum at about 120 to 150 µL/h and decreased at the higher rates. Therefore, in the present experiments, a subcutaneous tissue glucose monitoring system with a 15 mm long cellulose hollow fiber probe with 30 mm connecting tubing between the probe and a needle-type glucose sensor was applied with a perfusion rate 120 µL/h.

The recovery of glucose in the dialysate and relative response time of the cellulose follow fiber probe covered with and without MGC in vitro were determined. The percent recovery of glucose in the dialysate was 34.3 %. A rapid change occurred in the glucose concentrations by switching the medium from 100 to 200 mg/dL glucose, and the times to reach 90 % of the maximum and minimum values were 6.2 min and 8.1 min, respectively. These characteristics were not different from that for the original cellulose hollow fiber probe.

However, the possibility cannot be excluded that protein, fibrin and blood cells deposits might affect the permeability of the hollow fiber probe, resulting in the reduction of glucose recovery in the dialysate with time.

Figure 5 shows the time dependence of the glucose concentration as determined by the glucose monitoring system when the cellulose hollow fiber probe was incubated in saline solution containing plasma protein and albumin. In the case of the original cellulose hollow fiber probe, glucose concentration in the dialysate gradually decreased after 4 days of continuous dialysis. Moreover, the decrease occurred much faster with an increase in albumin concentration. On the contrary, during the 14-day

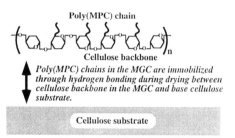

Figure 4. Schematic representation of coating state of MGC on cellulose substrate.

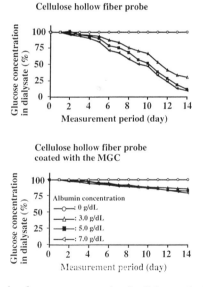

Figure 5. Changes in glucose concentration in dialysate during continuous dialysis in saline with albumin. The concentration of glucose was 200 mg/dL.

measurement periods, glucose recovery with the cellulose hollow fiber probe covered with MGC was fairly constant for 7 days even in the albumin solution, and then gradually decreased. The amount of albumin adsorbed on the cellulose membrane was larger than that covered with MGC. The difference in the stability between the original cellulose hollow fiber probe and that covered with MGC corresponds to the protein adsorption-resistance properties of the probes.

Continuous glucose monitoring in subcutaneous tissue using the microdialysis system. Figure 6 shows a representative result of the continuous glucose monitoring in subcutaneous tissue in a healthy subject using the microdialysis system with a MGC-modified probe. The subcutaneous tissue glucose concentrations could be monitored nicely for up to 7 days. On the other hand, with the original cellulose hollow fiber probe, subcutaneous tissue glucose concentrations could be measured for up to 4 days after which the ability significantly decreased on the 7th day.

As shown in Figure 7, with the original cellulose hollow fiber probe, a linear regression line between the apparent subcutaneous tissue glucose concentrations and blood glucose concentrations obtained after 4 days was not significantly different from that obtained on the 1st day, however, the regression line obtained on after 7 days was significantly different from the 1st day, demonstrating that apparent subcutaneous tissue glucose concentrations were lower than blood glucose concentrations. On the contrary, with the cellulose hollow fiber probe covered with MGC, a highly significant correlation between apparent subcutaneous tissue glucose concentrations and blood glucose concentrations was observed during the 7 days of monitoring. But, the regression line obtained after 10 or 14 days was significantly different from the 1st day. By introducing the *in vivo* calibration technique and correcting with fasting blood glucose levels every morning after 7 days continuous monitoring, subcutaneous tissue glucose concentrations measured by this monitoring system again correlated with glycemic excursions for 14 days.

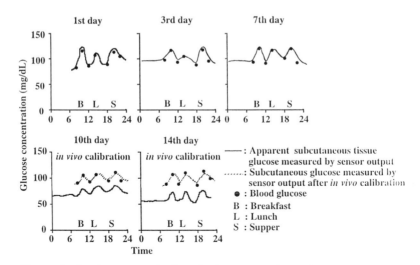

Figure 6. Continuous monitoring of subcutaneous tissue glucose concentrations in a healthy subject with a microdialysis sampling method using cellulose hollow fiber probe covered with MGC.

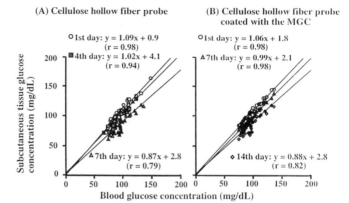

Figure 7. Linear regression analysis between subcutaneous tissue glucose concentrations measured by extracorporeal glucose monitoring system and blood glucose concentrations determined by glucose oxydase method in healthy subjects.

Transmission electron microscopic examination clearly demonstrated protein adsorption on the membrane surface after 4 days of continuous monitoring *in vivo* with the original cellulose hollow fiber probe(data not shown). On the other hand, protein adsorption on the probe covered with MGC was slightly observed during the 7 days of continuous application *in vivo*.

In conclusion, the biocompatibility of the cellulose hollow fiber probe covered with the MPC polymer is excellent. Therefore, the miniaturized extracorporeal monitoring system based on a microdialysis sampling method with the MPC polymer-modified cellulose hollow fiber probe combined a needle-type glucose sensor is stable and reliable for clinical applications.

Literature Cited

1. Shichiri, M.; Kawamori, R.; Yamasaki, Y.; Hakui, N.; Abe, H.; *Lancet,* **1982,** II, 1129.
2. Shichiri, M.; Fukushima, H.; Sakakida, M.; Kajiwara, K.; Hashiguchi, Y.; *Frontiers Med. Biol. Engng.,* **1991,** 3, 283.
3. Nakabayashi, N.; Ishihara, K.; *Macromol. Symp.,* **1996,** 101, 405.
4. Ishihara, K.; Aragaki, R.; Ueda, T.; Watanabe, A.; Nakabayashi, N.; *J. Biomed. Mater. Res.,* **1990,** 24, 1069.
5. Ishihara, K.; Ziats, N.P.; Tireney, B.P.; Nakabayashi, N.; Anderson, J.M.; *J. Biomed. Mater. Res.,* **1991,** 25, 1397.

6. Ishihara, K.; Oshida, H.; Ueda, T.; Endo, Y.; Watanabe, A.; Nakabayashi, N.; *J. Biomed. Mater. Res.*, **1992**, 26, 1543.
7. Ueda, T.; Ishihara, K.; Nakabayashi, N.; *J. Biomed. Mater. Res.*, **1995**, 29, 381.
8. Nishida, K.; Sakakida, M.; Ichinose, K.; Uemura, M.; Kajiwara, K.; Miyata, T.; Shichiri, M.; Ishihara, K.; Nakabayashi, N.; *Med. Progress Technol.*, **1995**, 21, 91.
9. Hashiguchi, Y.; Sakakida, M.; Nishida, K.; Uemura, T.; Kajiwara, K.; Shichiri, M.; *Diabetes Care*, **1994**, 17, 387.
10. Ishihara, K.; Nakabayashi, N.; Fukumoto, K.; Aoki, J.; *Biomaterials*, **1992**, 13, 145.
11. Ishihara, K.; Takayama, R.; Nakabayashi, N.; Fukumoto, K.; Aoki, J.; *Biomaterials*, **1992**, 13, 235.
12. Ishihara, K.; Miyazaki, H.; Kurosaki, T.; Nakabayashi, N.; *J. Biomed. Mater. Res.*, **1995**, 29, 181.
13. Ishihara, K.; Ueda, T.; Nakabayashi, N.; *Polym. J.*, **1990**, 22, 377.

Chapter 3

DNA-Modified Electrodes: Molecular Recognition and Electrochemical Response

Koji Nakano[1], Shinji Uchida, Yoshiharu Mitsuhashi, Yuji Fujita, Hiroaki Taira, and Mizuo Maeda[1]

Department of Chemical Science and Technology, Faculty of Engineering, Kyushu University, 6-10-1 Hakozaki, Higashi-ku, Fukuoka 812-81, Japan

A biosensor for the detection of the compounds that bind to DNA is reported. A chemical derivatization of terminal phosphate ends of DNA double strands (ds) with 2-hydroxyethyl disulfide was made to immobilize DNA onto Au surfaces via chemisorption. By taking advantages of the redox couple-mediated artificial ion-channel principle, the DNA-modified electrode was successfully applied for a bioaffinity sensor. For example, cyclic voltammograms (CV) of ferrocyanide/ferricyanide couple with the DNA-modified electrode gave the redox wave due to the reversible electrode reaction. The CV peak currents were significantly enhanced on adding quinacrine. The peak currents showed almost a linear relationship with the concentration of quinacrine in the range of $10^{-7} - 5 \times 10^{-7}$ M, and then saturated beyond the concentration of 8×10^{-7} M. CV responses toward a variety of DNA-binding substrates including quinacrine, acridine orange, safranin, spermine and spermidine showed a reasonable selectivity order according to the DNA-binding affinity of these compounds, while the response was quite small for methyl viologen which binds to ds DNAs through a nonspecific electrostatic interaction. The sensitivity was diminished when the immobilized ds DNAs were heat-denatured. The voltammetric response is primarily due to a "titration" of the immobilized ds DNAs by DNA-binding molecules. Thus the response-concentration profile for quinacrine was analyzed by using the Langmuir's isotherm. The apparent binding constant thus obtained was 1.3×10^{6} M^{-1} which agreed fairly well with that in literatures (1.5×10^{6} M^{-1}).

Molecular recognition is fundamental to most biochemical phenomena—it is crucial in enzymatic catalysis, immune reaction, DNA replication and in various processes occurring at cell membranes. Such an extraordinary property has been stimulating a considerable research interest in analytical applications. With using biological components or by mimicking of biological processes, a variety of biosensors have been reported (1). Enzyme-based electrochemical sensors are the most successful example. However, sensory application of DNA does not seem to have been fully studied.

The replication of DNA and its transcription to RNA is one of the most remarkable examples of molecular recognition—a single-stranded DNA (ss DNA) molecule seeks out for and hybridizes with its complementary strand. Because the RNA provides the

[1]Corresponding authors.

©1998 American Chemical Society

template for protein synthesis, the presence of base-sequence abnormalities such as a point mutation are known to cause inherited human diseases. Thus there has been an increasing demand to develop the systems capable of detecting the specific DNA sequences for use in diagnosis of such inherited diseases. The most promising way for this purpose is to use a DNA probe which has a sequence complementary to the target sequence in the DNA sample and hybridizes selectively with it under appropriate conditions. For the purpose of early diagnosis, on the other hand, DNA sensor is also a subject of considerable interest since it can provide a simple and rapid system. The basis of these sensors are the same as that of DNA probes: a ss DNA with a specific DNA sequence is immobilized on a solid support and hybridization with its target sequence in the DNA sample is detected by using voltammetric methods (2,3) or by coupling with mass-sensing devices (4).

In addition to the sequence specific hybridization between ssDNAs, double-stranded DNA (ds DNA) can interact with various molecules and ions (5,6). This process can be classified to four different modes—intercalation, electrostatic binding, binding to the minor or the major groove, and triple helix formation. Some polycyclic aromatic molecules bind to DNA by inserting themselves between the base pairs of the double helix. Metal ions bind to DNA by an electrostatic interaction with the phosphate group(s) or by a complex formation with the sugar moiety and the nucleic base. Some proteins such as repressors and promotors recognize a particular sequence of DNA and bind to that site through hydrogen bondings with the functional groups on the major groove. The fourth category of DNA-binding modes is that homopyrimidine oligodeoxynucleotide of 10 - 20 base pairs in length can form stable complexes in the major groove at homopurine sites with GC and AT complementarity. Therefore, ds DNA can be considered as a 'receptor' in the analytical sense.

As results of fundamental discoveries in biochemistry and pharmaceutical science, effects of DNA-binding of small molecules have been clarified. For example, the intercalation may lead to a useful result such as antibacterial or anticancer activity, but may also invoke undesirable responses such as carcinogenesis or mutagenesis. Natural or synthetic antitumor antibiotics which are categorized as a groove binder can exert their pharmacological effect in the same fashion. With these understanding one may notice the possibility of developing a bioaffinity sensor that responds selectively to DNA-binding substrates. The sensor would provide a certain mean for the evaluation of inherited, carcinogenic compounds. From this viewpoint, we have been investigating a bioaffinity sensor which comprises a DNA double helix as a receptive component (7-9). Our approach is as follows.

First, the terminal monophosphate end of a ds DNA was chemically modified with an organosulfur precursor. Then the whole ds DNA entity was immobilized on a Au electrode surface via chemisorption. Because a ds DNA has no such reactive functional groups as free amines or carboxylic acids, cross-linking reagents which are commonly used in enzyme immobilization seem to be ineffective. In the earlier study, we have characterized the immobilization chemistry by FT-IR spectroscopy and confirmed that the organosulfur precursor acts as an anchor to immobilize the whole DNA entity on a Au surface. Next, if one considers the sensory application of the DNA-modified electrode using an electrochemical technique, a problem arises because affinity reactions of ds DNAs are not directly linked to redox processes. In addition, the redox reactions of DNA bases are stoichiometric and do not involve chemical amplification. This is in a marked contrast to enzymatic reactions which often give potentiometrically or amperometrically measurable reaction products in an amplified manner. In order to overcome the difficulty in DNA, we took advantage of the principle of 'ion-channel sensor' which was proposed by Sugawara et al (10). Thus, the DNA-modified electrode was successfully applied for detecting quinacrine which is a well-known antimalarial drug (7).

In the present paper, cyclic voltammetric responses of the bioaffinity sensor toward a variety of DNA-binding substrates including quinacrine, acridine orange, safranin, spermine and spermidine were collected. The voltammetric response showed a reasonable selectivity order for these compounds, while the response was quite small for methyl viologen which caused a nonspecific electrostatic interaction with ds DNAs. The sensitivity of the sensor was diminished when the immobilized ds DNAs were heat-denatured. The voltammetric response is primarily due to a "titration" of the immobilized ds DNAs by DNA-binding molecules. Thus the response-concentration profile for quinacrine was analyzed by using the Langmuir's isotherm. The apparent binding constant thus obtained agreed fairly well with that in literatures. The result should support the validity of our approach. For the understanding of the voltammetric response, more detailed analysis of cyclic voltammograms based on convolution principle was also conducted.

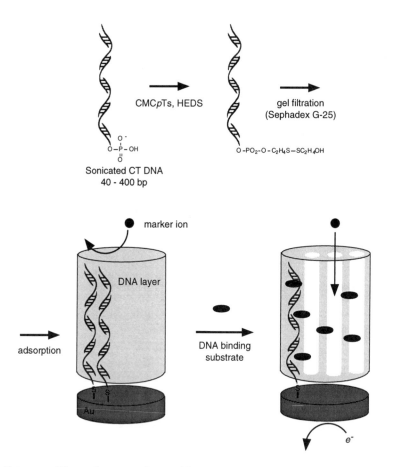

Scheme. Electrode preparation and DNA-dependent ion-channel mechanism.

Experimental

Chemicals. Acridine orange, quinacrine, safranine, spermine, spermidine, and methyl viologen were obtained from Aldrich Chemical Co. and used as received. The following reagents were of commercially available, the highest grade (Nacalai Tesque, Inc.) and used for the chemical derivatization of ds DNAs; 2-hydroxyethyl disulfide (HEDS), 1-cyclohexyl-3-(2-morpholinoethyl)carbodiimide metho-p-toluenesulfonate

(CMCpTs), dithiothreithol (DTT), and 5,5'-dithiobis-(2-nitrobenzoic acid) (DTNB). All other chemicals were guaranteed reagents and used without further purification.

Preparation of ds DNA fragments and chemical derivatization. Calf-thymus (CT) deoxyribonucleic acid sodium salt (Type I) was purchased from Sigma Chemical Co. The CT DNA was dissolved in 2 M KCl / 1 mM tris(hydroxymethy)aminomethane, pH 7.0 and the solution was subjected to sonication for fragmentation. Then the solution was purified by dialysis, followed by phenol extraction. The resulting ds DNA fragments (sonicated CT DNA) were obtained by ethanol precipitation. The sonicated CT DNA was applied to agarose gel electrophoresis and its molecular weight was determined; the molecular weight of the sonicated CT DNA prepared was ranging from 30 kDa to 260 kDa which was equivalent in base pairs (bp) from 40 to 400 (1 bp = 660 Da).

Sonicated CT DNA (10 mg) was reacted with HEDS (4.6 mg) in the presence of CMCpTs (0.4 g) in 0.4 cm^3 of MES buffer solution (0.04 M 2-(N-morpholino)ethanesulfonic acid–NaOH, pH 6.0) for 24 h at 25 °C to give a phosphodiester linkage between the terminal monophosphate ends of the DNA and the hydroxyl group of HEDS (*11, 12*). The reaction mixture was gel-filtrated on an NAP-10 column (Pharmacia), and the macromolecular fractions which displayed the characteristic absorption at 260 nm due to nucleic bases were collected. The combined fractions containing the modified DNA (0.3 mM in bp concentration) was stored at 5 °C and used directly for the surface modification of Au electrodes.

Determination of disulfide content. A 0.5 cm^3 portion of the disulfide modified DNA solution (5.2 mM in bp concentration) was added to a 0.5 cm^3 portion of 0.1 M phosphoric acid-NaOH buffer solution containing 20 μmol DTT and the mixture was allowed to stand for 1 hr at room temperature in order to reduce the disulfide into thiol groups. Then the excess amount of DTT in the reaction mixture was gel-separated on an NAP-10 column, and the macromolecular fractions were collected.

For the determination of the disulfide content of the modified DNA, Ellman's method was employed (*13*). DTNB was dissolved in 0.1 M phosphoric acid-NaOH buffer (pH 7.5) to give 10 mM of concentration. A 0.01 cm^3 portion of the solution was added to a 0.1 cm^3 portion of the collected DNA fractions. The concentration of thiophenolate anion which was released by the reduction of DTNB was determined spectrophotometrically (λ_{max}, 412 nm). The obtained concentration is equal to that of thiols in the sample solution. Comparing the concentration of thiols with the DNA concentration which was determined from UV absorption at 260 nm, we found that approximately 150 base pairs have 1 disulfide group. Since the DNA has some hundred of base pairs, each DNA strand seems to have one disulfide as its terminal group.

Electrode modification and electrochemical measurements. A gold disk electrode (diameter 1.6 mm, Bioanalytical Systems Co.) was polished to a mirror finish with alumina powders of successively finer grades (Buehler Co.), and was used as working electrode. The electrode was immersed into the disulfide-modified DNA solution for 24 h at 5 °C to modify the electrode surface. The modified electrode was then washed with and stored in TE buffer (10 mM Tris-HCl, 1 mM EDTA; pH 7.2) at 5 °C before and between uses. For denaturation of the immobilized ds DNAs, the modified electrode was first immersed into a hot (70 °C) aqueous solution of urea (6 M) for 15 min, and then successively treated with pure water which was cooled in an ice bath. The 'denatured' electrode was stored in the same manner as that described above.

Cyclic voltammetric (CV) measurements were made by using a Solartron Co., Model 1286 potentiostat. A conventional design of a three-electrode cell was used with a Pt plate (10 × 10 mm) counter electrode and a standard Ag/AgCl (3 M NaCl) reference electrode. Unless otherwise indicated, measurements were made at 25 mV s^{-1} on 5 mM each of ferrocyanide / ferricyanide solutions containing 0.01 M KCl. All experiments were made at a constant temperature of 25 ± 0.2 °C.

Results and Discussion

CV response to DNA-binding substrates. Most of DNA-binding substrates have a certain base-pair preference or base-sequence selectivity in their binding reaction with ds DNAs. This leads to an intrinsic variation of their binding constant depending on the base-pair content of the sample ds DNAs. The CT DNA contains a relatively even distribution of AT and GC base pairs, and thus it is preferable when examining compounds of unknown selectivity. This property is also favorable for use as a receptive component so that we used CT DNA in the present study.

Electrodes modified with ds DNA were examined by using cyclic voltammetry at 25 mV s^{-1} in a solution containing each 5 mM of Na$_4$[Fe(CN)$_6$] and Na$_3$[Fe(CN)$_6$] in 10 mM KCl. Under these conditions, bare Au electrodes give peak currents of 19 — 20 µA and the separation of the anodic and the cathodic peak potentials was 75 mV. The formal potential, taken as the average of the cathodic and the anodic peak potentials, was estimated to +0.207 V vs. Ag/AgCl. As reported previously, ds DNAs spread on the electrode surface significantly blocks the reversible redox reaction of the Fe(CN)$_6^{4-/3-}$ couple with the underlying electrode (7). CV measurements with using a number of independently prepared DNA-modified electrode showed that the magnitude of peak currents decreased at around a half relative to that measured on a bare Au electrode. In addition, an expansion of the potential difference was observed (ca. 300 mV). In contrast, treatment of a Au electrode with HEDS, which serves as an anchor for the attachment of ds DNA in the present immobilization chemistry, gave almost no effect on the electrode reaction of the redox species.

The CV profile of the solution species on the DNA-modified electrode significantly changes upon adding DNA-binding substrates. Typical CVs measured under various concentrations of quinacrine and spermine are shown in Figure 1. In case of quinacrine which is a well-known antimalarial drug, the peak current increases while the potential difference decreases with increasing concentration of the drug. Finally, both come close to those observed on a bare Au electrode, respectively. The peak currents showed almost a linear relationship with the concentration of quinacrine in the range of 10^{-7} — 5 × 10^{-7} M, and then saturated beyond the concentration of 8 × 10^{-7} M. Neither the bare Au electrode nor that treated with HEDS showed any response to quinacrine. The binding of the cationic quinacrine molecule to the immobilized polyanionic host should reduce the negative charge on the electrode surface, resulting in the enhancement of the electrode reaction of the solution species. Figure 1 also shows CVs taken in the presence of spermine which binds to the major groove of ds DNAs. As can be seen in this figure, the present system shows CV response to spermine in the same manner as that of quinacrine. Such concentration-dependent changes in CV profiles were also observed for other intercalators including acridine orange and safranine, and a groove binder, spermidine (date are not shown).

The CV response of the present system was almost reversible to changes in substrate concentration but the electrode itself became less stable with the elapse of time. After a series of measurements, the electrode recovered the magnitude of CVs' peak currents of the initial state within a 5 % deviation by exposing to TE buffer solution. On the other hand, the long-time stability of the electrode was varied for each electrode preparation of CV experiments. Unfortunately, we can not make it clear how long the electrode can be used with a good reproducibility. We would like to claim at the present stage that the electrode is of course not a single-use device and, even in the worst case, repeated uses (3 times) within 24 h give reproducible responses toward DNA-binding substrates.

Selectivity of the CV response. As noted in the preceding section, addition of a DNA-binding substrate into the measurement solution makes the electrode reaction of the Fe(CN)$_6^{4-/3-}$ redox couple more feasible. This results in changes in the magnitudes of peak currents (i_p) and as well as in the separation of peak potentials (ΔE_p). The former should be proportional to the concentration of the redox couple in the vicinity of the electrode surface and the latter is associated with the rate of electron transfer

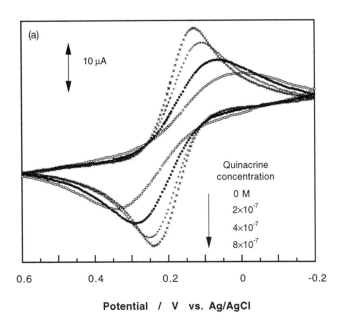

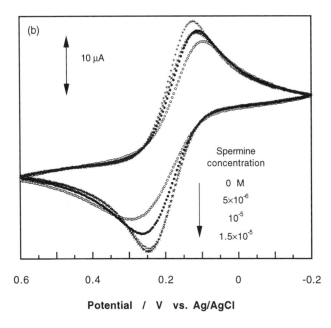

Figure 1. Effects of (a) quinacrine and (b) spermine on current-potential curves for the DNA-modified electrode; at 25 °C; scan rate, 25 mV s^{-1}; [K$_4$[Fe(CN)$_6$]] = [K$_3$[Fe(CN)$_6$]] = 5 mM, [KCl] = 10 mM.

between the solution species and the electrode. These two parameters should be considered for the detailed discussion of effects of the affinity reaction occurring at the electrode surface on the electrode reaction of the solution species. First, we will discuss the CV response by the present system to the DNA-binding substrates on the basis of changes in the peak current (Δi_p) because i_p is the most suitable characteristic of CV in the practical sense.

Figure 2 summarizes the CV responses by the present system to DNA-binding substrates. For intercalators and groove binders, a steep rise followed by a saturation of the concentration-response is commonly observed, suggesting a saturation binding to the immobilized ds DNA. If one compares the specific concentration which gives a 50 % response in Δi_p for each substrates, a selectivity order of quinacrine $\approx$ acridine orange > spermine > spermidine > safranin can be estimated. The binding constants measured in aqueous media for the affinity reaction with ds DNA are as follows: quinacrine, 1.5×10^6 M^{-1} (38 mM NaCl) (14); acridine orange, $(1.2 \pm 0.3) \times 10^6$ M^{-1} (ionic strength, 0.02) (15); spermine, 2.1×10^5 M^{-1} (94 mM NaCl) (16); spermidine, 5.8×10^4 M^{-1} (94 mM NaCl) (16). The selectivity order by the present sensor system seems consistent with the binding affinity of the substrates with ds DNA. That is, the selectivity of the system is primarily determined by the binding affinity of the substrate with DNA duplex. We have recently reported metal-ion responses of the present system; the electrode reaction of the Fe(CN)$_6^{4-/3-}$ on the DNA-modified electrode was significantly enhanced on adding metal ions, showing a selectivity order of Mg^{2+} > Ca^{2+}, Ba^{2+} >> Na$^+$, K$^+$ (9). These observations support a potential feasibility of the present system for application to a bioaffinity sensor which responds selectively to DNA-binding substrates.

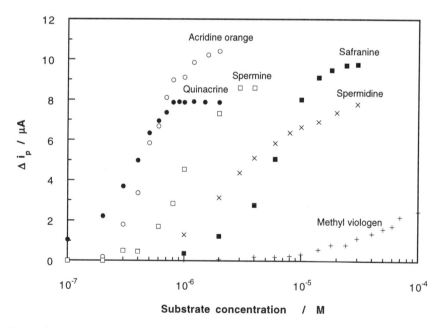

Figure 2. DNA-binding substrate dependent changes on the cathodic peak current of the cyclic voltammograms taken similarly to those in Figure 1.

In contrast, methylviologen which has a binding constant of 1.8×10^5 M^{-1} (1.5 mM cacodylate, pH 6) (*17*) to ds DNA gives much more insensitive CV response than that of other substrates. Although we can not fully understand the anomalous CV response for methyl viologen, one possible explanation may be the difference in the binding mode to ds DNA; methyl viologen binds to the exterior of DNA helix *via* electrostatic forces. Intercalation of a planar molecule between the stacked bases of ds DNA brings about some unwinding and enlongation of DNA double helix. Binding of such polyamines such as spermine and spermidine in the negatively charged major groove of the double helix by coulombic interactions with ribose phosphate groups as well as by possible formation of hydrogen bondings with the nucleic bases induces collapse of nucleic acids. As far as we know, such marked structural changes of ds DNAs through the electrostatic binding of cationic molecules to the exterior of ds DNAs have not been reported. If such a conformational changes of ds DNAs are associated with the binding process which takes place on the modified electrode surface, it should also enhance or suppress the electrode reaction of the $Fe(CN)_6^{4-/3-}$ redox couple with the underlying electrode. We think that the low sensitivity to methyl viologen in the present CV response may arise from the guest-induced conformational changes of DNA molecules immobilized on the electrode surface.

The concentration-response curves for acridine orange and safranin gave somewhat larger value of Δi_p at saturation than those of other substrates. These two compounds are redox active and thus there is a possibility that their DNA-complexes act as an electron mediator for the electrode reaction of the solution species. Voltammetric studies for their redox properties on the DNA-modified electrode are now in progress.

Thermal denaturation experiments and the CV responses. CV responses were also studied for the modified electrodes which were treated by the thermal denaturation procedure of the DNA duplex to give single-stranded form. As a result, the difference in selectivity diminished almost completely and the magnitude of Δi_p lowered for all the substrates. The denatured ss DNA still has polyanioic charges and thus, binding of substrates having cationic charges is to occur through coulombic interactions. This should reduce the negative charge on the electrode surface, resulting in the enhancement of the electrode reaction of the solution species. Disapperance of the selectivity order upon the denaturation procedure is reasonable because no specific interaction other than electrostatic forces is expected for the binding of substrates with ss DNAs. These results also suggest that the selectivity of the present system is primarily determined by the binding affinity of the substrate with the DNA duplex.

Binding isotherm estimated from CV responses. The concentration-dependent changes of i_p suggest saturation binding of these substrates to the immobilized ds DNAs. We assume a 1 : 1 binding reaction between a substrate molecule and a binding site composed of two adjacent base pairs on the DNA duplex. It has been reported that intercalation binding of quinacrine has maximum saturation at one bound molecule per two base pairs. In other words, after one site is occupied, binding at neighboring site is excluded. So at saturation of binding, the concentration of the occupied site should be the same as that of the free site. Now the binding constant can be simply expressed by the reciprocal of concentration of free quinacrine at saturation. The quinacrine concentration at saturation, on the other hand, can be read as 6.2×10^{-7} M from Figure 2. Then we can obtain apparent binding constant to be 1.6×10^6 M^{-1}. According to the literature, quinacrine has binding constant of 1.5×10^6 M^{-1} in the similar conditions to the present ones in terms of temperature and concentration of coexisting salts. This fair agreement between two values led us to quantify the binding isotherm based on the calibration data.

For the Langmuir-type binding isotherm we take three assumptions: (i) only a 1 : 1 binding is to occur non-cooperatively between a substrate molecule (M) and a physicochemically equivalent binding site ($S_{DNA, s}$) locating on the solid electrode surface, and (ii) the binding reaction does not significantly alter the bulk substrate concentration. For the estimation of the ratio of bound to free substrate (*b*) , (iii) the normalized values of i_p in CV responses at any substrate concentration are used. These assumptions gives the binding equilibrium represented as follows.

$$M + S_{DNA,s} \rightleftharpoons MS_{DNA,s} \qquad K = \frac{[MS_{DNA,S}]}{[M][S_{DNA,S}]}$$

$$b = \frac{[MS_{DNA,S}]}{[M]+[S_{DNA,S}]} = \frac{K[M]}{1+K[M]} = \frac{i_p - i_{p,\min}}{i_{p,\max} - i_{p,\min}} \qquad (1)$$

The latter equation can be arranged,

$$b^{-1} - 1 = \frac{1}{K[M]} \qquad (2)$$

or,

$$\frac{1-b}{b} = K[S] \qquad (3)$$

Plots of (b^{-1}-1) vs reciprocal of concentration for quinacrine are shown in Figure 3. These data are correlated by a straight line and the apparent binding constant, K, is obtained to be 1.32×10^6 M^{-1}. In case of acridine orange, the use of Eq. (3) gave a more reliable result; the logarithmic plots followed a straight line and the apparent binding constant was determined as 1.95×10^6 M^{-1}. For these two compounds, interaction with ds DNA was studied in a homogeneous, aqueous solution using a spectrophotometric method and binding constants of 1.5×10^6 M^{-1} and $(1.2 \pm 0.3) \times 10^6$ M^{-1} were reported, respectively. Although there is no experimental evidence that the normalized i_p is equivalent to the ratio of bound to free substrate, the binding constants determined from CV responses agreed fairly well with that determined by a spectrophotometric method. It is a bit surprising that the binding reaction between substrate molecules and ds DNAs which are immobilized on an electrode surface occurred in the similar manner as that in a homogeneous system.

For groove binders, on the other hand, such simple treatments based on either Eq. (2) or Eq. (3) did not give satisfactorily results; plots showed a considerable curvature, suggesting some cooperative effect in their binding with ds DNA. The McGee-von Hippel equation in which the binding site is taken into account is desirable for the analysis of the binding reaction of ds DNA with small molecules. Because it was impossible to determine the intrinsic amount of substrate either bound to ds DNA or free in solution only through CV measurements, we gave up the more detailed analysis of CV data.

Convolution analysis for CVs and quinacrine response. As noted in the preceding section, addition of a DNA-binding substrate into the measurement solution makes the electrode reaction of the $Fe(CN)_6^{4-/3-}$ redox couple on the DNA-modified electrode more feasible. In addition to changes in the magnitude of i_p, we observed decrease of ΔE_p which is associated with the rate of electron transfer between the redox couple and the electrode. For all compounds tested in the present study, the peak wave shifts to more positive potentials in the case of a reduction and to more negative potentials in the case of an oxidation with increasing concentration of the substrate. This means that the affinity reaction between the substrates and ds DNAs immobilized on the electrode surface leads to enhancement of the apparent heterogenous electron transfer

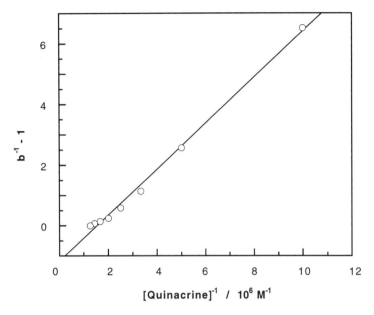

Figure 3. Plots of $(b^{-1} - 1)$ vs reciprocal of concentration for the cyclic voltammetric response to quinacrine on the DNA-modified electrode.

rate of the solution species on the DNA-modified electrode. At the present stage, however, the close understanding of the redox process occuring on the modified electrode surface is yet incomplete. Such experimental complications as the intrinsic surface area of the modified electrode and mass-transfer process through the ds DNA layer of the solution species as well as the counter ion make the further data processing quite difficult. Although the following discussion is still in the insufficient stage, we will describe the concentration dependence of ΔE_p by using the quinacrine response as an example.

The voltammetric $i\text{-}E$ curves were first transformed into the steady-state voltammetric curves (*18*). This transformation makes use of the convolution principle based on a numerical integration technique on a computer and corrects the diffusion effect in CVs of the solution species. The corrected voltammetric curves for the oxidation waves of the $Fe(CN)_6^{4-/3-}$ redox couple are shown in Figure 4. The normalized values of the anodic current at any quinacrine concentration were plotted against the electrode potential, and the apparent values of the half-wave potential ($E_{1/2}$) were determined through a linear regression. Variation of $E_{1/2}$ of the redox couple on the concentration of quinacrine is shown in Figure 5. The concentration-dependent profile of $E_{1/2}$ fairly resembles to that of i_p. That is, it decreases with increasing concentration of quinacrine in the range of 10^{-7} — 5×10^{-7} M, and saturated beyond the concentration of 6×10^{-7} M. The saturated value of $E_{1/2}$ was almost the same as that on the bare Au electrode. Binding of cationic quinacrine molecules with the polyanionic ds DNAs should reduce the electrostatic repulsion between the anionic redox couple and ds DNAs. This should lead to the enhancement of the mass-transfer process of the redox

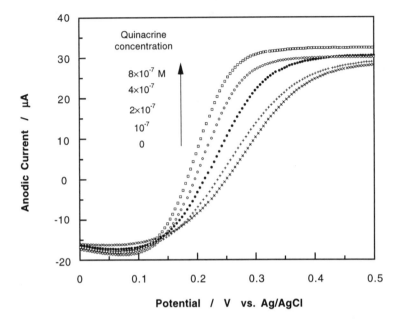

Figure 4. Convoluted voltammograms of for the electrode reaction of $[Fe(CN)_6]^{4-/3-}$ on the DNA-modified electrode in the presence of various concentration of quinacrine; experimental conditions are the same as those in Figure 1.

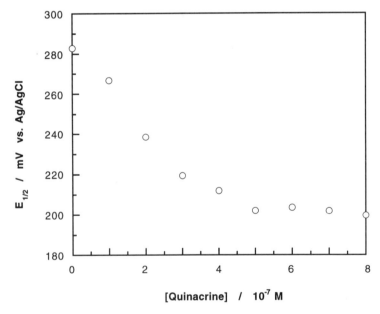

Figure 5. Plots of $E_{1/2}$ vs quinacrine concentration for the voltammograms shown in Figure 4.

couple in the vicinity of the electrode surface which is covered with ds DNAs and may enhance the apparent heterogeneous electron transfer rate of the redox couple on the modified electrode surface. Anyhow, there are some problems remaining unsolved in the close understanding of the redox reaction on the DNA-modified electrode surface and these are the subject for a further study.

In addition, the saturated values of $E_{1/2}$ as well as i_p were almost the same as those on a bare Au electrode. The ds DNA molecules are, in a general sense, considered to be an insulator. Immobilization of ds DNA on an electrode surface should reduce the effective surface area of the electrode. Thus, our observation is somewhat imcomprehensible. Recently, energy- and electron-transfer between the stacked aromatic heterocycles in the interior of a DNA helix is of a research interest. Barton *et al.* (*19*) claimed a long-range photoinduced electron-transfer through a DNA helix while Harriman *et al.* (*20*) reported nagative results. In case of an electrochemical system, possibility of electronic conduction through ds DNAs which were immobilized on an electrode surface was suggested. Future subjects lie in this region.

Acknowledgements

We thank Professor K. Takehara of Kyushu University for use of a computer program of the convolution procedure. This work was supported in part by a Grant-in-Aid for Scientific Research from the Ministry of Education, Science and Culture of Japan. Financial supports by The Salt Science Research Foundation (to M.M.) and by The Fukuoka Industry, Science and Technology Foundation (to K.N.) are also acknowledged.

Literature Cited

1. *Biosensors. A Practical Approach*; Cass, A. E. G., Ed., IRL Press, New York, 1990.
2. Hashimoto, K.; Miwa, K.; Goto, M.; Ishimori, Y. *Supramol. Chem.* **1993**, *2*, 265.
3. Millan, K. M.; Mikkelsen, S. R. *Anal. Chem.* **1993**, *65*, 2317.
4. Ebersole, R. C.; Miller, J. A.; Moran, J. R.; Ward, M. D. *J. Am. Chem. Soc.* **1990**, *112*, 3239.
5. Wilson, W. D.; Jones, R. L. In *Intercalation Chemistry*; Whittingham, M. S., Jacobson, A. J., Eds., Academic Press, New York, 1982, pp 445-501.
6. Sigel, H. *Chem. Soc., Rev.,* **1993**, 255.
7. Maeda, M.; Mitsuhashi, Y.; Nakano, K.; Takagi, M. *Anal. Sci.,* **1992**, *8*, 83.
8. Maeda, M.; Nakano, K.; Takagi, M. In *Diagnostic Biosensor Polymers*; Usmani, A. M., Akmal, N., Eds., American Chemical Society, Washingthon, DC, 1994; Vol. 556.
9. Maeda, M.; Nakano, K.; Uchida, S.; Takagi, M. *Chem. Lett.,* **1994**, 1805.
10. Sugawara, M.; Kojima, K.; Sazawa, H.; Umezawa, Y. *Anal. Chem.,* **1987**, *59*, 2842.
11. Rickwood, D. *Biochim. Biophys. Acta,* **1972**, *269*, 47.
12. Nicola, N. A.; Kristjansson, J. K. R.; Fasman, G. D. *Arch. Biochem. Biophys.,* **1979**, *193*, 204.
13. Habeeb, A. F. S. A. *Methods Enzymol.,* **1972**, *25*, 457.
14. Wilson, W. D.; Lopp, I. G. *Biopolymers,* **1979**, *18*, 3025.
15. Armstrong, R. W.; Kurucsev, T.; Strauss, U. P. *J. Am. Chem. Soc.,* **1970**, *92*, 3174.
16. Schneider, H.-J.; Blatter, T. *Angew. Chem. Int Ed. Engl.,* **1992**, *31*, 1207.
17. Fromherz, P.; Rieger, B. *J. Am. Chem. Soc.,* **1986**, *108*, 5361.
18. Bard, A. J.; Faulker, L. R. *Electrochemical Methods. Fundamentals and Applications*, Wiely, New York, 1980.
19. Murphy, C. J.; Arkin, M. R.; Jenkins, Y; Ghatlia, N. D.; Bossmann, S. H.; Turro, N. J.; Barton, J. D. *Science,* **1993**, *262*, 1025.
20. Brun, A. M.; Harriman, A. *J. Am. Chem. Soc.,* **1994**, *116*, 10383.

Chapter 4

Amperometric Biosensors Using an Enzyme-Containing Polyion Complex

Fumio Mizutani, Soichi Yabuki, Yukari Sato, and Yoshiki Hirata

National Institute of Bioscience and Human-Technology, 1-1 Higashi, Tsukuba, Ibaraki 305, Japan

Amperometric enzyme electrodes for L-lactic acid and ethanol were prepared by immobilizing lactate oxidase and alcohol oxidase into polyion complex membranes, respectively: three kinds of aqueous solutions, i.e., a poly(4-styrenesulfonate) solution, an enzyme solution and a poly-L-lysine solution, were successively placed on a glassy carbon electrode and dried. The anodic current (1 V *vs.* Ag/AgCl) of each enzyme electrode increased after the addition of the corresponding analyte, owing to the electrolytic oxidation of the hydrogen peroxide produced through the oxidase-catalyzed reaction in the membrane. The membrane showed permselectivity based on the solute size with the molecular weight cut-off of ca. 100 g/mole. The permselectivity was effective in reducing the interferencial response as compared to the response to the analyte: the permeation of interferents such as L-ascorbic acid, uric acid and acetaminophen, was restricted, whereas the analyte permeated easily to undergo the enzymatic reaction. Another ion-complex based electrode, an enzyme monolayer-attached electrode was produced through the co-adsorption of the enzyme (lactate oxidase) and poly-L-lysine onto a mercaptoalkanoic acid-modified gold electrode. The enzyme molecules were entrapped in a matrix consisting of the mercaptoalkanoic acid and the cationic polymer. The resulting lactate-sensing electrode show a rapid response (100% response time, less than 0.5 s) to the analyte.

Amperometric biosensors with immobilized enzymes have attracted increasing interest in the last two decades. Ideally, the specificity of enzymes combined with high sensitivity of electrochemical signal transduction should result in enzyme electrode capable of measuring the concentration of the enzyme substrate in a medium containing a mixture of other compounds.

In conventional enzyme electrodes, at least two membranes are placed on the electrode surface: one is an enzyme membrane and the other is a semipermeable

©1998 American Chemical Society

membrane that eliminates electrochemical interferents such as L-ascorbic acid, uric acid and acetaminophen. However, the required semipermeable membrane complicates the structure of enzyme electrode and also increases the cost. Therefore, the immobilization of enzyme in a semipermeable membrane seems to be of particular interest. Since the molecular weights of the interferents are higher than 150 g/mole, semipermeable membranes with molecular weight cut-off of < 150 g/mole are useful for the preparation of enzyme electrodes to the substrates with relatively low molecular weights, e.g. L-lactate- and ethanol-sensing electrodes (molecular weights of the analytes, 90 g/mole and 46 g/mole, respectively).

Recently we have found that enzymes could be immobilized into a polyion complex membrane prepared from poly-L-lysine and poly(4-styrenesulfonate) and that the membrane showed permselectivity based on the solute size with the molecular cut-off of around 100 g/mole (1). Enzyme molecules were entrapped into the ionically-crosslinked complex, as illustrated in Figure 1a, and the crosslink topology was effective for the selective permeation of small molecules (2-4). In this paper, we report the preparation and the use of two kinds of enzyme electrodes, i.e., L-lactate- and ethanol-sensing ones, based on the poly-L-lysine/poly(4-styrenesulfonate)-complex membrane. The enzyme electrode for L-lactic acid (1,5,6) and ethanol have been proved to be suitable for the selective measurement of corresponding analyte in sera and/or drinks.

Further, we have prepared ultra-thin layers containing enzymes (7,8) by combining the method for the enzyme-immobilization mentioned above (1) with that for preparing a monolayer of poly-L-lysine on a negatively-charged, mercaptoalkanoic acid-modified gold surface (9). That is, a mixture of an enzyme and poly-L-lysine and was placed on a mercaptoalkanoic acid-modified gold electrode, so as to form a mercaptoalkanoic acid/poly-L-lysine-matrix with the immobilized enzyme, as shown in Figure 1b. Preliminary results for the ultra-thin layer-based electrode are also reported.

Experimental

Reagents. The enzymes used were lactate oxidase (LOx; EC number, not assigned; from *Pediococcus* sp., 40 U mg^{-1}, Sigma, St. Louis, MO), alcohol oxidase (AOx; EC 1.1.3.13, from *Hansenula* sp., 12 U mg^{-1}, Sigma) and peroxidase (EC 1.11.1.7, from horseradish, 200 U mg^{-1}, Boehringer Mannheim, Indianapolis, IN). Poly-L-lysine hydrobromide (molecular weight, 90,000 g/mole) and lithium L-lactate were obtained from Sigma, and poly(sodium 4-styrenesulfonate) (molecular weight, 70,000 g/mole), from Aldrich (Milwaukee, WI). Other reagents were of analytical reagent grade (Nacalai Tesque, Kyoto). Deionized, doubly-distilled water was used throughout. F-Kits (Boeheringer Mannheim) were used for the spectrophotometric measurement of L-lactic acid and ethanol. The kit for L-lactic acid uses the enzyme pair of lactate dehydrogenase (EC 1.1.1.27) and glutamate-pyruvate transaminase (EC 2.6.1.2) and that for ethanol, alcohol dehydrogenase (EC 1.1.1.1) and aldehyde dehydrogenase (EC 1.2.1.5). Human sera were purchased from Sigma and ICN (Costa Mesa, CA).

Poly-L-Lysine/Poly(4-styrenesulfonate)-Complex-Based Electrodes. A glassy

48

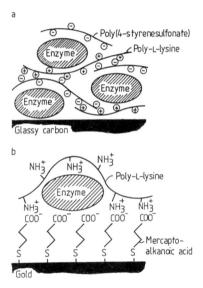

Figure 1. Schematic illustrations of (a) the enzyme-containing poly-L-lysine/poly(4-styrenesulfonate)-complex membrane and (b) the enzyme ultra-thin layer consisting of an enzyme-containing mercaptoalkanoic acid/poly-L-lysine matrix.

carbon disc (diameter 3 mm; Bioanalytical Systems, West Lafyyette, IN) was used as the base electrode: glassy carbon is a cheap material and useful for the anodic detection of the hydrogen peroxide produced through oxidase-catalyzed reaction, although a rather high operating potential (around 1 V vs. Ag/AgCl) is required (10,11). The electrode was polished with a 0.05-μm alumina slurry, rinsed with water, sonicated in water for 2 min, and then dried. Then, three kinds of aqueous solutions [a poly(4-styrenesulfonate) solution (6 μl, 25 mM in the monomer unit), an enzyme (LOx or AOx) solution (10 μl, concentration of each enzyme, 2%(w/v)) and a poly-L-lysine solution (10 μl, 20 mM in the monomer unit)] were successively placed on the electrode surface, and the electrode was allowed to dry at room temperature for 4 h. The thickness of each membrane was estimated to be several microns by using a stylus-type apparatus (Talystep, Tayler-Hobson, Leicester). The poly-L-lysine/poly(4-styrenesulfonate)-complex-based electrode thus prepared was stored in 0.1 M potassium phosphate buffer (pH 7.0) at 4°C when not in use.

Enzyme Ultra-thin Layer-Coated Electrode. A gold disc electrode, 1.6 mm in diameter (Bioanalytical Systems), was polished with a 0.05-μm alumina slurry, rinsed with water, sonicated in water for 2 min, then cleaned by an electrochemical oxidation/reduction treatment (+1.5 to −0.3 V vs. Ag/AgCl with a potential weep rate of 0.1 V/s) in 0.05 M H_2SO_4 for 30 min. The layer of 3-mercaptopropionic acid (MPA) was prepared onto the gold surface by immersing the electrode in an ethanol solution containing the mercaptoalkanoic acid (1 mM) for 1 h at room temperature. After washing the electrode with ethanol and then water, it was immersed an aqueous solution containing LOx (0.5 mM) and poly-L-lysine (1 mM per lysine residue) for 24 h at 4°C. The electrode was then thoroughly washed with 0.1 M potassium phosphate buffer (pH 7.0). The enzyme electrode (**E 1**) thus prepared was stored in 0.1 M potassium phosphate buffer (pH 7.0) at 4°C when not in use.

Other LOx-based electrodes were prepared as follows. A MPA/LOx-electrode (**E 2**) was made by immersing the MPA-modified electrode in a 0.5 mM LOx solution for 24 h at 4°C. A poly-L-lysine/LOx-bilayer electrode (**E 3**) was prepared by applying the procedure of electrostatic layer-by-layer adsorption (12); first, positively-charged poly-L-lysine was adsorbed onto the MPA-modified electrode (9) by immersing it in an aqueous solution containing the polycation (1 mM in the monomer unit) for 24 h at 4°C, then, the electrode was washed with water and immersed into a 0.5 mM LOx solution for 24 h at 4°C as to form a monolayer of negatively-charged LOx (pI, 4.6) (13) onto the poly-L-lysine layer (12).

Measurements. A potentiostat (HA-502, Hokuto Denko, Tokyo) was used in a three-electrode configuration for amperometric measurements: the enzyme electrode, an Ag/AgCl reference electrode (Bioanalytical Systems) and a platinum auxiliary electrode were immersed in 20 ml of a test solution. The test solution was 0.1 M potassium phosphate buffer whose pH was 7.7 for testing the LOx-based electrodes and 7.0 for the AOx-based electrodes. The potential of the poly-L-lysine/poly(4-styrenesulfonate)-complex-based electrode was set at 1.0 V vs. Ag/AgCl, and that of the enzyme ultra-thin layer-coated electrode, at 0.7 V vs. Ag/AgCl.

The enzyme activities were measured by using the peroxidase-phenol-4-aminoantipyrine chromogenic system (14).

Poly-L-Lysine/Poly(4-styrenesulfonate)-Complex-Based Electrodes

Enzyme Activity and Permselectivity of the Polyion Complex Membrane. A light yellow precipitate consisting of an enzyme (LOx or AOx)-containing polyion complex was formed by the addition of poly-L-lysine to the mixture of poly(4-styrenesulfonate) and an enzyme. After drying, a layer consisting of the precipitated product was formed on the electrode surface. The layer was not removed from the electrode surface in a vigorously-stirred aqueous solution and showed the enzymatic activities of 55 mU cm^{-2} for the LOx-based system and 25 mU cm^{-2} for the AOx-based one. Therefore, a layer of a poly-L-lysine/poly(4-styrenesulfonate) complex containing enzyme molecules is considered to be formed as illustrated in Figure 1a.

The solute permeability dependence as a function of molecular weight for the LOx-immobilized polyion complex membrane was examined as follows. Anodic current responses on the polyion complex-based electrode and a bare glassy carbon electrode were measured for a variety of electroactive species whose molecular weights ranged from 30 g/mole (hydrazine) to 665 g/mole (NADH). The ratio of current response on the polyion complex-based electrode to that on the bare glassy carbon electrode for each species was plotted against its molecular weight, as shown in Fig. 2. The ratio was higher than 0.5 for the electroactive species having a molecular weight less than 100 g/mole, and sharply decreased with increase in the molecular weight over 100 g/mole. Thus, the present polyion complex-based membrane proved to be useful for the preparation of L-lactate- and alcohol-sensing electrodes. Such an ultrafilter is suitable for the elimination of electrochemical interferents such as L-ascorbic acid, uric acid and acetaminophen, whereas each anlayte permeates with a high flux.

L-Lactate-Sensing Electrode. Figure 3 shows the response-time curves for the LOx/polyion complex-based electrode to the successive addition of L-lactic acid and that of the three kinds of electroactive species, i.e., L-ascorbic acid, uric acid and acetaminophen. After the addition of L-lactic acid, the electrode current increased and reached a steady-state within 5 s: the fast electrode response was obtained owing to the simple layer structure of the present electrode. The steady-state current increase was proportional to the L-lactate concentration up to 0.3 mM, and a significant increase in the current with increasing the L-lactate concentration was obtained in the range 0.3 to 1.0 mM. The detection limit was 0.1 µM. The relative standard deviation for ten successive measurement of 0.05 mM L-lactic acid was within 2%. As shown in Figure 3, the addition of each electroactive species brought about a discernible increase in the electrode current. This shows that these species interfere in the measurement of L-lactic acid. However, the ratio of response for each interferent to that for the same concentration of L-lactic acid was relatively small, owing to the restriction of transport of the interferents in the polyion complex membrane (Figure 2); the ratios were 0.25, 0.40 and 0.55 for L-ascorbic acid, uric acid and acetaminophen, respectively.

The concentrations of L-lactic acid, L-ascorbic acid, uric acid and acetaminophen in

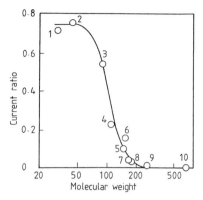

Figure 2. Effect of the molecular weight of oxidizable species on the ratio of current increase after the addition of the species for the polyion complex-based electrode to that for a bare electrode. The species employed were (1) hydrazine, (2) hydrogen peroxide, (3) oxalic acid, (4) hydroquinone, (5) acetaminophen, (6) dithiothreitol, (7) uric acid, (8) L-ascorbic acid, (9) 1,1-ferrocenedicarboxylic acid, and (10) NADH. The concentrations of the species were 5 mM for oxalic acid and 0.1 mM for others, and the potentials applied to the electrodes were 1.3 V *vs.* Ag/AgCl for oxalic acid and 1.0 V *vs.* Ag/AgCl for others. Each ratio was averaged over the measurements by the use of three different pairs of polyion complex-based and bare electrodes.

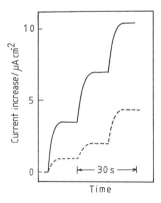

Figure 3. Current-time curves for the LOD/polyion complex-based electrode obtained on increasing the L-lactate concentration in 0.05 mM step (——) and the successive addition of L-ascorbic acid, uric acid and acetaminophen (-----). The concentration of each electroactive species was 0.05 mM.

normal sera are around 2, 0.05, 0.3 and 0 mM, respectively. The concentrations of L-lactic acid and L-ascorbic acid in sour milk samples are ca. 50 mM and less than 2 mM, respectively. For measuring L-lactic acid in the sera and sour milk with the LOx/polyion complex-based electrode, the current changes caused by the interferents are expected to be ca. 7% and 1% for that from the analyte, respectively. Table I shows the results for the determination of L-lactic acid in human sera and sour milk from the current increase on the enzyme electrode. Each sample (0.5 ml for serum and 20 μl for sour milk) was used without any pretreatment. The results were compared with those given by the F-kit method. The agreement was excellent for both sera and sour milk. These results demonstrate that the LOx/polyion complex-based electrode is useful for the simple and rapid determination of L-lactic acid in real samples.

Table I. Comparison of results obtained for L-lactic acid in human sera and sour milk by different methods

Sample	L-Lactate concentration (mM)	
	proposed method	F-kit method
Serum 1	2.03	2.02
Serum 2	1.44	1.36
Serum 3	2.80	2.86
Serum 4	1.75	1.67
Sour milk 1	63.8	61.4
Sour milk 2	74.6	73.0
Sour milk 3	53.3	53.6
Sour milk 4	87.7	86.1
Sour milk 5	72.4	74.4

The long-term stability of the LOx/polyion complex-based electrode was examined by determining 0.1 mM L-lactic acid 10 times daily for 12 weeks. The average value of the electrode response for the 10 measurements did not reduce for 4 weeks at all. The electrode response gradually decreased after 4 weeks, but it was still measurable until the end of 12th week: the current response on the 84th day was 30% of the initial value.

The LOx/polyion complex-based thus showed high performance characteristics such as rapid response, wide linear range and high stability. The enzyme electrode has been proved to be useful as the detector in an FIA system for measuring L-lactic acid in sera (6).

Alcohol-Sensing Electrode. Figure 4 shows the current-time curve on the AOx/polyion complex-based electrode to the successive addition of methanol, ethanol and L-ascorbic acid. After the addition of each alcohol, the electrode current increased and reached a steady state within 5 s. For ethanol, the electrode response was proportional to the substrate concentration up to 0.4 mM and the detection limit was 0.2 μM. The ratio of response for L-ascorbic acid to that for the same concentration of ethanol was 0.6. Such an AOx-based electrode exhibited much larger current response for methanol than ethanol. The current response was proportional to the methanol

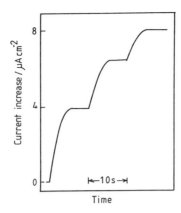

Figure 4. Current-time curve for the successive addition of 0.05 mM methanol, 0.1 mM ethanol and 0.1 mM L-ascorbic acid on the AOx/polyion complex-based electrode.

concentration only up to 0.1mM, but six times larger than the ethanol response in the linear concentration region.

Table II gives the results for the determination of ethanol in wines and sakes. The results compared with those obtained by the F-kit method. The agreement is excellent, which showed that the AOx/polyion complex-based electrode is useful for determining ethanol in alcoholic drinks.

Table II. Comparison of results obtained for ethanol in alcoholic drinks by different methods

Sample	Ethanol concentration (M)	
	proposed method	F-kit method
Wine 1	1.92	1.96
Wine 2	2.09	2.01
Sake 1	2.69	2.58
Sake 2	2.29	2.30

The long-term stability of the AOx/polyion complex-based electrode was then examined by determining 0.05 mM ethanol 10 times daily for two weeks. The average value of the electrode response for the 10 measurements did not reduce for one week, but the electrode response gradually decreased after one week to be 50% of the initial value at the end of the second week, as shown in Figure 5. The rather poor stability of the AOx/polyion complex-based electrode would be caused by the rapid deactivation AOx. However, this would be circumvented by the simplicity of the preparative procedure of the enzyme electrode.

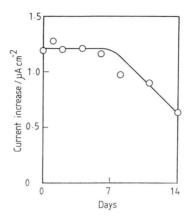

Figure 5. Long-term stability of the AOx/polyion complex-based electrode. The electrode response to 0.05 mM ethanol was measured 10 times a day. The average value of the 10 measurements was plotted against the number of days after the preparation of the electrode.

Enzyme Ultra-thin Layer-Coated Electrode

Electrode Response. Figure 6 shows current-time curves on **E 1-3** after the addition of 0.1 mM L-lactic acid. The current on each electrode increased immediately after the addition of L-lactic acid and reached a steady state within 0.5 s. As each enzyme layer is extremely thin, the substrate added is expected to diffuse quickly through the layer so as to produce hydrogen peroxide through the LOx reaction, which would result in a rapid response. The current response for **E 1** was proportional to the concentration of L-lactic acid up to 0.2 mM. The detection limit was 0.01 mM, and the relative standard deviation for 10 successive measurement of 0.1 mM L-lactic acid was 2%.

As shown in Figure 6, **E 1** gave much larger current response than other electrodes. For **E 2** and **E 3**, the current response to L-lactic acid is considered to be brought about by the LOx molecules adsorbed on the MPA-modified and poly-L-lysine-coated electrodes, respectively. The nonspecific adsorption of LOx would limit the number of the enzyme molecules on the electrode, as compared to the case of enzyme entrapment in the MPA/poly-L-lysine matrix.

Surface Concentration of LOx Molecules. The numbers of LOx molecules on the three kinds of electrodes were estimated as follows. The potential of each electrode was set at -1.1 V *vs.* Ag/AgCl for 30 min for the reductive desorption of thiol compound from the gold surface (*15*), which accompanied the dissolution of enzyme molecules into the solution (*7*). This experiment was carried out by employing a glassy carbon disc (diameter 3 mm) as an auxiliary electrode, in order to avoid the deposition of the dissolved enzyme molecules on the auxiliary electrode. Negatively charged enzyme molecules such as LOx molecules have been known to deposit onto a platinum

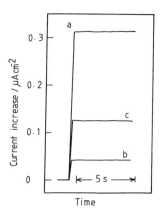

Figure 6. Current responses to 0.1 mM L-lactic acid for (a) E 1, (b) E 2 and (c) E 3.

anode (*16*), whereas such a deposition of enzyme was suppressed by employing glassy carbon as an anode material (*17*). The LOx activities in the test solutions after the cathodic treatment of **E 1-3** were 0.25, 0.05 and 0.10 mU, respectively. By assuming that the enzyme activity remained unchanged through the immobilization and desorption processes, the amounts of LOx molecules that had been attached on the surfaces of **E 1-3** were estimated to be 4×10^{-12}, 8×10^{-13}, and 2×10^{-12} mol cm^{-2}, respectively. These results indicate that the magnitude of current response to L-lactic acid depends on the surface concentration of LOx, and that LOx molecules seems to form a monolayer on each electrode. The diameter of LOx molecules is 6-7 nm, and the maximum coverage of LOx monolayer is estimated to be 5×10^{-12} mol cm^{-2} (ignoring roughness factor). The MPA/poly-L-lysine matrix is useful for preparing an LOx monolayer with a high surface coverage.

Thus an ultra-thin layer containing LOx was prepared by the co-adsorption of the enzyme and poly-L-lysine onto a negatively-charged, mercaptoalkanoic acid-modified electrode. The present method can be applied to the immobilization of a variety of proteins. We have already prepared a glucose-sensing electrode by immobilizing glucose oxidase (*6*).

Conclusion

Amperometric enzyme electrodes for L-lactic acid and alcohol were prepared by immobilizing LOx and AOx into poly-L-lysine/poly(4-styrenesulfonate)-complex membranes, respectively. The membrane showed permselectivity based on the solute size with the molecular weight cut-off of ca. 100, which was effective in reducing the interferential response as compared to the response to the analyte for specific applications

An ultra-thin enzyme layer was prepared by the co-adsorption of poly-L-lysine and the enzyme (LOx) onto a mercaptoalkanoic acid-modified gold electrode. The enzyme molecules were entrapped in a crosslinked matrix consisting of the carboxylic acid on the electrode and the cationic polymer.

Literature Cited

(1) Mizutani, F.; Yabuki, S.; Hirata, Y. *Anal. Chim. Acta*, **1995**, *314*, 233-239.
(2) Bixler, H. J.; Michaels, A. S. *Encl. Polym. Sci. Eng.*, **1969**, *10*, 765-780.
(3) Smid, J; Fish, D. *Encl. Polym. Sci. Eng.*, **1988**, *11*, 720-739.
(4) Markley, L. L.; Bixler, H. J.; Cross, R. A. *J. Biomed. Mater. Res.*, **1968**, *2*, 145-155.
(5) Mizutani, F.; Yabuki, S.; Hirata, Y. *Denki Kagaku*, **1995**, *63*, 1100-1105.
(6) Mizutani, F.; Yabuki, S.; Hirata, Y. *Talanta*, **1996**, *43*, 1815-1820.
(7) Mizutani, F.; Sato, Y.; Yabuki, S.; Hirata, Y. *Chem. Lett.*, **1996**, 251-252.
(8) Mizutani, F.; Sato, Y.; Hirata, Y.; Yabuki, S. *Denki Kagaku*, **1996**, *64*, 1266-1268.
(9) Jordan, C. E.; Frey, B. L.; Kornguth, S.; Corn, R. M. *Langmuir*, **1994**, *10*, 3642-3648.
(10) Gorton, L. *Anal. Chim. Acta*, **1985**, *178*, 247-253.
(11) Mizutani, F.; Yabuki, S.; Katsura, T. *Anal. Chim. Acta*, **1993**, *274*, 201-207.
(12) Luvov, Y.; Ariga, K.; Kunitake, T. *Chem. Lett.*, **1994**, 2323-2326.
(13) Enzyme Catalog, Toyo Jozo, Tokyo, **1982**, T-13.
(14) Bauminger, B. B. *J. Clin. Pathol.*, **1974**, *27*, 1015-1017.
(15) Walczak, M. M.; Popenoe, D. D.; Deinhemmer, R. S.; Lamp, B. D.; Chang, C.; Poter, M. D. *Langmuir*, **1991**, *7*, 2687-2693.
(16) Aizawa, M.; Chiba, T.; Shinohara, H. *Nippon Kagaku Kaishi*, **1987**, 2210-2213.
(17) Yabuki, S.; Mizutani, F. *Denki Kagaku*, **1995**, *63*, 654-659.

Chapter 5

High-Linearity Glucose Enzyme Electrodes for Food Industries: Preparation by a Plasma Polymerization Technique

Mehmet Mutlu[1,3], Selma Mutlu [2,3], Ismail H. Boyaci[1], Burçak Alp[1], and Erhan Piskin[2,3]

[1]Food Engineering Department, [2]Chemical Engineering Department, and [3]Bioengineering Division, Hacettepe University, 06532 Ankara, Turkey

A plasma polymerisation technique was used to prepare high linearity amperometric enzyme electrodes. Two different types of recognition layer for glucose based on oxidases were studied. The first type of recognition layer was prepared in sandwich form with polyethersulphone (PES) and polycarbonate (PC) membranes which were used as inner and outer membranes, respectively. Glucose oxidase was entrapped between these two membranes. The outer surface of PC membranes were further modified by deposition of hexamethyldisiloxane (HMDS) in a glow-discharge reactor to improve linearity. A new type of recognition layer in the form of a single membrane was prepared from cellulose acetate (CA). One surface of the cellulose acetate membrane was modified by deposition of amylamine (AmA) by plasma polymerisation to substitute amino groups on the membrane surface for glucose oxidase immobilisation. Both types of recognition layers were placed separately on a probe type oxygen electrode system with Pt working electrode and Ag/AgCl reference electrode which was polarised at 650 mV. Linearity of these enzyme electrodes were investigated. The linearity of the sandwich type electrode was extended to 600 mM glucose by surface modification of the outer membrane by glow-discharge deposition of hexamethyldisiloxane. The linearity of the single layer enzyme electrode was found to be 1000 mM glucose.

Food engineering amongst the fields now being transformed by continually increasing level of automation. Whereas the objective in other sectors of industry is simply to increase efficiency, here considerations of system theory or safety demand a high level of automation. Either the processes are too complex and require multifunctional control with feedback, or an analysis of the safety requirements shows the necessity for a certain degree of redundancy in the safety measures, and for elimination of human error as a risk factor (1).

The operating concentration range of the glucose biosensor based on glucose oxidase (GOD), when coupled with electrochemical detection of hydrogen peroxide, is narrow, however, and dilution of samples is always required. Food samples (soft and energy-supplying drinks, fruit juices, etc.) generally have higher glucose concentrations than clinical samples (up to 500 mM).

The enzyme electrodes, a subgroup of biosensors, are amongst the most widely studied electrodes for food industries. There are numerous researches on improvement of oxidases based electrodes which are used for the determination of corresponding substrate in diluted and undiluted food stuff. The main interest of an amperometric transducer is the linear relationship existing between the output current and the analyte concentration, in contrast to potentiometric detection, which leads to a logarithmic calibration. One of the main reasons restricting the use of biosensors is the low linear calibration range which is limited by the Michaelis constant of the enzyme (K_m). The concentration range of the glucose under determination may be extended in two ways: by dilution of samples or by the design of membranes possessing high diffusion limitation with respect to the substrate (2,3).

In this paper, we focused on extending linearity of glucose enzyme electrodes by a novel technique so called "plasma polymerisation". Plasma polymerisation was selected as the method of surface modification either for polymer deposition or enzyme immobilisation. Some satisfactory applications of plasma polymerisation for surface modification have been elaborated by our group (4-6) and others (7,8). The outer membrane of the conventional sandwich type recognition layer was modified in a glow-discharge reactor with an organosilane, hexamethyldisiloxane. We also prepared a new type of single membrane recognition layer which gathers all the functions on itself. Here, we present the performance of recognition layers on the basis of electrode linearity.

Preparation of the recognition layers

Materials. Low density polycarbonate membranes (LDPCM) with nominal pore size 0.01 μm (rated by manufacturer, $1*10^8$ pores/cm^2) were supplied by Nuclepore (Pleasenton, CA). Amylamine (AmA) was obtained from BDH (UK). Hexamethyldisiloxane (HMDS) was obtained from Wacker (Germany). Glucose oxidase (E.C.1.1.3.4. from Aspergillus Niger: 292 IU mg^{-1} protein) was purchased from Sigma (UK). Polyethersulphone (PES) were supplied by ICI (UK). Cellulose acetate (CA) (39.8 % acetyl content) was obtained from Aldrich (Germany). Gluteraldehyde (25% aqueous solution), bovine serum albumin (fraction V), glucose, solvents, buffer components and other standart reagents were obtained from Sigma (UK).

Electrode. A probe type electrode system which was produced in our laboratories was used. The Pt electrode was polarised at 650 mV (vs. Ag/AgCl) for hydrogen peroxide detection, the meter was linked to a data acquisition system of a computer. The electrode consisted of a central 2 mm diameter platinum working electrode with an outer 12 mm diameter silver ring as the reference.

Sandwich Type Recognition Layer (SRL). The outer membrane of the sandwich type recognition layer was a track-etched low density polycarbonate membrane (LDPCM) with 0.01 μm pore size was purchased from Nuclepore. The outer surface of this membrane was modified by glow-discharge treatment. In this technique, HMDS was deposited by plasma polymerization on the LDPCM in a glow-discharge reactor system. In a typical glow-discharge treatment, two membranes attached to each other (outer surfaces looking to plasma medium) were placed in the middle of the reactor. The reactor was evacuated to 10^{-3}-10^{-4} mbar. The monomer (i.e. HMDS) allowed to flow through the reactor at a flow rate of 30 ml/min. The LDPC membranes were exposed to plasma medium for 5 min at a discharge power of 10 W. The details of the plasma modification for surface modification were given in our previous studies (6).

The inner membrane of the sandwich type recognition layer i.e. Polyethersulphone (PES) membranes were prepared by a solvent casting technique (9). A 6% (w/v) solution of PES in a mixture of 2-methoxyethanol and DMF (25% + 75%, v/v) was casted on a glass plate. The polymer film (the thickness : 25 μm) was then casted by evaporation of the solvents in a conditioned room (relative humidity: 80% and temperature: 20°C). The membranes were kept in vacuum desiccator until use.

The enzyme layer was prepared by crosslinking of glucose oxidase (GOD) with glutaraldehyde, in which GOD was diluted (co-crosslinked) by mixing with bovine serum albumin (BSA). In a typical procedure, 30 mg of GOD was mixed with a 200 mg BSA in 1 ml of buffer solution. A 6 μl volume of this mixture was then rapidly mixed with 3 μl of glutaraldehyde (5%, v/v) in a Gilson pipette tip and transferred between the LDPC and PES membranes. This sandwich was then compressed between two glass slides and held under hand pressure for 5 minutes. The glass slides were pried apart and the membrane/enzyme laminate was left to dry at room temperature for a further 5 min., washed with buffer and then placed over the tip of probe.

Single Membrane Recognition Layer (SMRL). The main body of the single recognition layer made up of cellulose acetate (CA) membrane. Cellulose acetate (CA with 39.8% acetyl content) membranes (with a thickness of 25 μm) were also prepared by solvent casting technique. The composition of the casting solution was 2 % CA (w/v) in acetone. The precipitation was achieved by evaporation of acetone at room temperature (20°C) with 80% relative humidity. The membranes were kept in vacuum desiccator until use.

The cellulose acetate membranes then were treated in a glow discharge reactor by amylamine (AmA) monomer to incorporate amino groups on the surface. Only one side of the cellulose acetate membranes were modified by plasma polymerisation. The monomer (i.e. AmA) allowed to flow through the reactor at a flow rate of 30 ml/min. The CA membranes were exposed to plasma medium for 5 min at a discharge power of 5 W.

The glow-discharge treated cellulose acetate membranes, PlzP(AmA)-CA (25 cm^2 surface area), were incubated in a reaction vessel containing 10 ml of 12.5% (v/v) glutaraldehyde aqueous solution for 24 hr at 37 °C. The glutaraldehyde activated membranes were then washed with distilled water to remove any unbound glutaraldehyde. An enzyme solution was prepared with the glucose oxidase (GOD) content of 10 mg in 10ml of buffer solution and these membranes were then placed into the beakers and left to interact for 24 hrs at 37 °C. In order to prevent stagnant film formation on the PlzP(AmA)-CA membrane surfaces, activation media were continuously mixed during the coupling process. At the end of the incubation period the membranes were removed and washed with a sequence of the following buffers; 0.1 M acetate buffer (pH:4), 0.1 M borate buffer (pH:8), 0.1 M acetate buffer (pH:4) and finally 0.5 M NaCl to remove non-specifically bound protein. The general preparation procedure of single membrane recognition layer was given schematically in Figure 1.

The position of recognition layer on the tip of the electrode was studied and the performance of enzyme electrode was reported in our previous paper (M. Mutlu, S. Mutlu, B. Alp, İ.H. Boyacı, E. Pişkin, "Preparation of a Single Layer Enzyme

Electrode by Plasma Polymerisation Technique", in "Plasma Treatments and Deposition of Polymers", Ricardo D'Agustino (ed.), NATO ASI Series, Kluwer Academic Publishers, Dordrecht, in press).

Linearity Tests of the Glucose Electrodes

The linearity of the sandwich type recognition layer (SRL) and single membrane recognition layer (SMRL) were tested with glucose. The experimental set-up were shown in Figure 2. Both types of recognition layers were placed at the tip of the probe type electrode, separately. The mass transfer barriers and reaction steps of each type recognition layer were sketched in Figure 3. For the tests carried out with single membrane recognition layer, the enzyme bound surface of the membrane was faced with the solution to achieve higher linearity (M. Mutlu, S. Mutlu, B. Alp, İ.H. Boyacı, E. Pişkin, "Preparation of a Single Layer Enzyme Electrode by Plasma Polymerisation Technique", in "Plasma Treatments and Deposition of Polymers", Ricardo D'Agustino (ed.), NATO ASI Series, Kluwer Academic Publishers, Dordrecht, in press). In these tests, the test solution was an isotonic buffer solution (pH: 7.4) containing 0.0528 mol/l Na_2HPO_4, 0.0156 mol/l NaH_2PO_2, 0.0051 mol/l NaCl and 0.00015 mol/l EDTA. In the tests, for both type of recognition layers, the enzyme electrode was first allowed to stabilize for 5-10 min in the buffer solution. Then, 5 μl of an aqueous solution of glucose was added to achieve a 10 mM glucose concentration in the medium. The response was measured. This procedure was repeated several times, without changing the solution but adding enough glucose solution to increase its concentration by 10 mM each time, until the deviation from linearity was observed.

Results and Discussion

Linearity of the enzyme electrodes with sandwich type recognition layer: Sandwich type amperometric enzyme electrodes are amongst the most widely studied electrodes for the last three decades. The outer membrane mainly controls the substrate (i.e., glucose) diffusion and therefore provides the linearity. This membrane should also allow sufficient oxygen transfer which is vital for the enzymatic reaction. In the middle of the sandwich laminate, an enzyme layer is located which reacts with the substrate by using oxygen and produces electrochemically detectable species (e.g., H_2O_2). Note that the enzyme is used in the immobilized (crosslinked) form in order to prevent the enzyme leakage and to increase the enzymatic stability. To control the enzymatic conversion rate, the enzyme concentration in this layer is controlled by adding diluting agents. In the case of the sandwich type glucose electrode constructed in this study, GOD molecules were co-crosslinked with BSA molecules in a proper ratio (i.e., 30/200) to control the enzymatic activity. The effect of crosslinker density and enzyme/BSA ratio to the mass transfer and kinetic behaviour of the enzyme electrode were studied by our group (10, S. Mutlu, M. Mutlu, E. Pişkin, A Kinetic Approach To Oxidases Based Enzyme Electrodes : The Effect of Enzyme Layer Formation on the Response Time, *J. Biochemical Engineering*, in press). In any type glucose electrode, the reaction product is hydrogen peroxide. In sandwich type recognition layer, H_2O_2 molecules pass through the inner membrane, reach to the electrode surface, and create the electrode response by oxidation. The most important property of this inner membrane is its' permselectivity. This membrane should only permit the H_2O_2 molecules, but not the interfering electroactive compounds which may come from the food samples to be analysed. PES membranes fulfilled all the requirements of a permselective membrane (6).

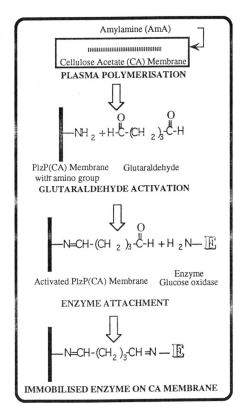

Figure 1. Preparation steps of single membrane recognition layer, SMRL

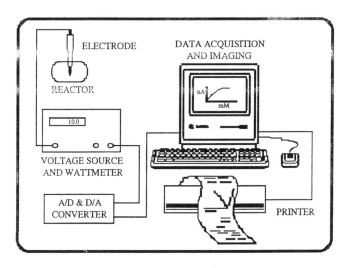

Figure 2. Experimental Set-up

In this study, we prepared sandwich type of laminates by using low density polycarbonate based membranes with 0.01 μm pore size as the outer membrane. In order to increase the linear range of the commercially available PC membranes by decreasing the pore size, we attempted to use a new technique, which is so-called "glow-discharge treatment". Note that by this technique, its possible to change the surface properties of a membrane without significantly changing the bulk structure (11). We deposited hexamethlydisiloxane (HMDS) on the LDPC membranes by plasma polymerization in a glow-discharge reactor. The effect of the HMDS deposition on the electrode linearity is given in Figure 4.

Figure 4. clearly shows that the linearity of the sandwich type enzyme electrode constructed with plain low density poycarbonate membrane with 0.01 μm pore size was found to be 60 mM glucose. The effect of plasma treatment on the linearity (mM glucose) and the response (nA) on LDPCM are also given in Figure 4. The deposition of organosilane (i.e. HMDS) has provided a 10 fold greater linear range (600 mM glucose). The extention of the linearity by plasma treatment can be explained by creation of a film on membrane and inside walls of the pores. So, polymer deposition was reduced the pore diameter, but was not apperently totally blocked all available pores. Therefore, the limitation of the mass transfer of the substrate (glucose) through outer membrane extends the linear range of the electrode.

These results can be compared with our previous studies which were carried out with high density polycarbonate membranes (HDPCM) with the same pore size. The pore density of HDPCM was $8*10^8$ pore/cm^2 which was rated by manufacturer, Poretics Co. The linearity of the electrode with plain HDPCM as outer membrane was found to be 10 mM of glucose (6).

It can be concluded that, to decrease mass transfer area of outer membrane by means of reducing pore density per unit area, or reducing the total mass tranfer area by polymer deposition, results in the extention of the linearity of the enzyme electrode.

Linearity of the enzyme electrode with single membrane recognition layer: Permselective behaviour of different polymeric membranes (cellulose acetate, polycarbonate, polyurethane, polyetersulphone) were investigated for a group of different electroactive compounds such as; acetic acid, ascorbic acid, citric acid, ethyl alcohol, lactic acid, malic acid, methyl alcohol, sodium benzoate, oxalic acid, orthophosphoric acid and tartaric acid, which may found in food stuff (12). PES and cellulose acetate membranes fulfilled all the requirements of a permselective membrane. Therefore, in this part of the study, cellulose acetate membrane is selected as the main matrix of the single membrane recognition layer.

Glow-discharge treatment was selected as the method of surface modification of the membranes for enzyme immobilisation. The main point for selection of this approach is that, it is a simple derivatization procedure without applying any complicated chemistry and without using any harmful chemicals which may damage both on enzyme activity and on the membrane integrity, to achieve high immobilised enzyme yield and activity.

In this study, one side of the cellulose acetate membranes were first treated in a glow discharge reactor by using amylamine (AmA) plasma, to substitute amino groups on the membrane surface, which would be available for further modification. The glow discharge conditions were optimised to achieve the highest amount of amino groups on the surface (13). The optimal conditions were as follows; exposure time: 5 min, discharge power: 5 W, monomer flow rate: 30 ml/min. The presence of the free amino groups on the membrane were determined by acetilation of the amino groups on the membrane with radiolabelled [1-14C] acetic anhydride followed by

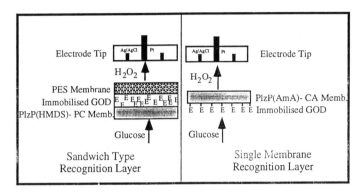

Figure 3. The mass transfer barriers and reaction steps in the sandwich type and the single membrane recognition layers

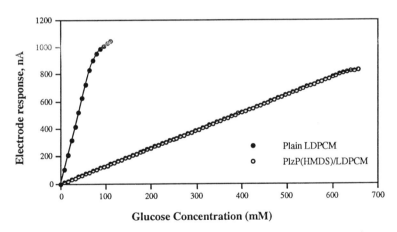

Glucose Concentration (mM)

Figure 4. Calibration curve of the sandwich type enzyme electrode

64

heterogeneous scintillation counting of the respective membranes. The details of this technique were given in our previous studies (14).

As described at the second step of Figure 1., glutaraldehyde was used as a cross-linker for enzyme immobilisation for the later step. The amino-group modified CA membranes were treated with the excess amount of crosslinking agent, glutaraldehyde, to activate the surfaces for enzyme immobilisation. Than the enzyme, glucose oxidase, was allowed to react with the aldehyde groups on the membrane surface. The immobilisation conditions were optimised to achieve the maximum yield and activity (6). Therefore, the formation of the recognition layer of the enzyme electrode has been completed.

The calibration curve of the cellulose acetate based single membrane recognition layer was given in Figure 5. It is clearly observed that, the linearity of the electrode is extending upto 1000 mM glucose concentration. This value is quite reasonable for the electrodes to be used directly in food analysis, process lines, reactor control, etc.

Single layer enzyme electrodes also facilitates the transport of the electroactive compound through membrane layer faster than sandwich type enzyme electrodes. In this case, there is only one mass transfer barrier for H_2O_2 to reach the electrode surface. The studies concerning the response time of electrode is still underinvestigation by our group (M. Mutlu, S. Mutlu, B. Alp, İ.H. Boyacı, E. Pişkin, "Preparation of a Single Layer Enzyme Electrode by Plasma Polymerisation Technique", in "Plasma Treatments and Deposition of Polymers", Ricardo D'Agustino (ed.), NATO ASI Series, Kluwer Academic Publishers, Dordrecht, in press).

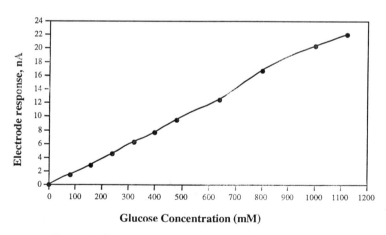

Figure 5. Calibration curve of the single layer electrode

Conclusion

We have presented a summary of experiments aimed to extend the linearity of the electrodes for direct measurement of glucose in food stuff. Plasma polymerisation technique was employed to modify surfaces of low density polycarbonate membranes and cellulose acetate membranes. Low density polycarbonate membranes were used as the outer membrane of the classical sandwich type recognition layer. Plasma modification of low density polycarbonate membranes by organosilanes were resulted as the increase in diffusion limitation of substrate through membrane and the linearity of the electrode was extended about 10 fold. The single membrane recognition layer was prepared by glow discharge to substitute amino groups for enzyme immobilisation . In this case cellulose acetate membranes which is highly selective to electroactive compounds but H_2O_2 were used. The plasma polymerisation treated cellulose acetate membranes with immobilised enzyme on itself were used as a recognition layer of the enzyme electrode. The linearity of electrode were found to be 1000 mM glucose. Finally, we point out that these results augment the usability of plasma polymerisation technique to prepare high linearity enzyme electrodes for food industry.

Acknowledgement

This research is partly supported by United Nations Developing Programme (UNDP) and Turkish State Planning Organisation (DPT) joint project. Project No: TUR-92-004.

References

1. Camman, K.; Lemke, U.; Rohen, A.; Sander, J.; Wilken, H.; Winter, B., *Angew. Chem. Int. Ed. Engl.* **1991**, *30*, 516-539.
2. Amine, A.; Patriarche, G.J.; Marrazza, G.; Mascini, M., *Analytica Chim. Acta* **1991**, *242*, 91-98.
3. Kanapieniene, J.J.; Dedinaite, A.A.; Laurinavicius, V.-S.A., *Sensors and Actuators B* **1992**, *10*, 37-40.
4. Mutlu, M.; Ercan, M.T.; Pişkin E.,*Clinical Materials* **1989**, *4*, 61-76.
5. Mutlu, M.; Pişkin, E., *Medical and Biological Engineering and Computing* **1990**, *28*, 232-236.
6. Mutlu, S.; Mutlu, M.; Vadgama, P.; Pişkin E., *American Chemical Society (ACS) Symposium Series* **1994**, *556*, 71-83.
7. Yasuda, H., *J. Polym. Sci. Macromol. Reviews* **1981**, *16*, 199.
8. Chawla, A.S., In *Polymeric Biomaterials*; Pişkin, E. and Hoffman, A.S., Eds.; NATO ASI Series, E 106, 1986, pp 231-243.
9. Battersby, C. M.; Vadgama, P., *Analytica Chimica Acta* **1986**, *vol 183*, 59-66.
10. Mutlu, M.; Mutlu, S., *J. Biotechnology Techniques* **1995**, *9(4)*, 277-282.
11. Pişkin, E., *J. Biomater. Sci. Polymer Edn.* **1992**, *vol 4*, 45-60.
12. Alp, B.; Mutlu, S.; Pişkin, E.; Mutlu, M., *Proc. Tenth Forum for Applied Biotechnology* **1996**, *vol 1*, 1615-1618.
13. Mutlu, S., *Ph. D. Dissertation*, Hacettepe University, Ankara, 1994.
14. Mutlu, M; Mutlu, S.; Rosenberg, M.F.; Kane, J.; Jones, M.N.; Vadgama, P., *J. Materials Chemistry* **1991**, *1*, 447-450.

Chapter 6

Micro, Planar-Form Lactate Biosensor for Biomedical Applications

Sayed A. Marzouk[1,4], Vasile V. Cosofret[1,5], Richard P. Buck[1], Hua Yang[2], Wayne E. Cascio[2], and Saad S. M. Hassan[3]

[1]Department of Chemistry, University of North Carolina, Chapel Hill, NC 27599–3290
[2]Department of Medicine, University of North Carolina, Chapel Hill, NC 27599-7075
[3]Department of Chemistry, Ain Shams University, Cairo, Egypt

Lactic acid is an important metabolic product formed during anaerobic glycolysis. There are no continuous methods to measure the extracellular lactate accumulation occurring in the absence of myocardial perfusion. Two forms of biosensors have been constructed. The first miniature version fulfills the operational requirements, such as flexibility, wide linear response range, high selectivity, and fast response time. This biosensor is based on immobilized lactate oxidase with anodic detection of the produced H_2O_2. An inner layer, which allows selective detection of hydrogen peroxide, is formed by electropolymerization of m-phenylenediamine. A diffusional barrier of polyurethane greatly increases the linear response range. The second miniature sensor uses an anode made of tetrathiofulvalene-tetracyanoquinodimethane (TTF-TCNQ) charge transfer complex. The former sensor was used successfully under reduced ambient oxygen tension ($PO_2 > 24$mm Hg); while the second can operate in the absence of oxygen. The applications are measurements of lactate in the ischemic rabbit papillary muscle under no-flow conditions.

Connection of Lactate Generation with Ischemic Events

Deprivation of molecular O_2 from respiring cells shifts the production of high energy phosphates from oxidative phosphorylation to glycolysis. An end product of anaerobic glycolysis is lactic acid (1,2). This is the reason that blood lactate level is an important metabolic indicator for presence and extent of anaerobic glycolysis. The latter occurs normally during vigorous exercise as the energy demands of skeletal muscle exceed the

[4]Current address: Department of Chemistry, Ain Shams University, Cairo, Egypt
[5]Current address: Instrumentation Laboratories, Lexington, MA, 02713

©1998 American Chemical Society

supply of O_2 substrate. Anaerobic glycolysis contributes to the pathophysiology of hypoxia, anoxia or ischemia. Under these conditions excessive production and accumulation of intracellular H^+, in part from lactic acid production, results in a decrease of intracellular pH where the fall in intracellular acidosis contributes to reduced myocardial contractility and altered impulse propagation.

Because of the pH dependence of many cellular enzyme systems, the cell has evolved numerous ion channels, pumps, exchangers and transporters to maintain an optimal intracellular pH. Transsarcolemmal transport of lactate-anion and diffusion of lactic acid are believed to be important mechanisms regulating intracellular pH under these conditions. However, the interaction of these various cellular mechanisms for the regulation of intracellular pH under metabolic stress is not completely understood. A complete understanding of these processes is lacking, in part, because an experimental method to measure extracellular lactate, continuously and directly, with sufficient temporal and spatial resolution under no-flow conditions did not exist, prior to our work.

Biosensors as a Reliable Tool. The previously described miniature version of a conventional lactate sensor was applied to the measurement of lactate accumulation in the papillary muscle of a rabbit heart suffering hypoxia, a mild ischemic event with low, but non-zero oxygen pressure (3). Three configurations of Au-based amperometric electrodes were fabricated by a photolithographic technique on flexible polyimide (Kapton) foils. All sensors were based on an immobilized lactate oxidase with detection of the hydrogen peroxide produced and oxidized at a platinum-electroplated Au base electrode at 0.5V vs Ag/AgCl. An inner electropolymeric layer was used to reject interferences by easily oxidized molecules. A diffusion controlling outer layer of cast polyurethane film extended the response range as required by the generation of lactate during ischemic conditions. These sensors have high selectivity, good operational stability, good accuracy and precision (average recovery = 102.3 ± 0.4% for control sera), fast response time (t_{95} = 20 s) and high upper limit of the linear dynamic range (25-80 mM, with sensitivity of 1.7-0.4 nA/mM at PO_2 = 15 mmHg). Fig. 1 shows the cross sectional construction of the anode.

    Subsequently, fabrication, characterization and application of a miniaturized amperometric lactate biosensor was described that supplements or may replace earlier designs. It is third generation type, based on crosslinked lactate oxidase and tetrathiafulvalene-tetracyanoquinodimethane (TTF-TCNQ) charge transfer complex. The sensor was developed for continuous quantitative monitoring of lactate accumulation in ischemic myocardium *under severe depletion of oxygen*. The sensor was evaluated *in vitro* at an applied potential of 0.15 V vs Ag/AgCl; it proved to combine all the performance characteristics desired for the present application, such as proper response in absence of oxygen, good operational stability, good accuracy and precision (103.5 ± 1.2 %), adequate response time ($t_{95\%}$ = 80 s), and wide linear dynamic range up to 27 mM (r=0.9998) in N_2-saturated solutions and at 37 °C. The prepared sensors (n = 12) showed sensitivity of 380 ± 94 nA/mM, and a background current of 240 ± 50 nA. The lower limit of detection is 0.4 ± 0.15 mM with S/N ratio equal to 3. Results obtained for direct lactate monitoring in ischemic rabbit papillary muscle under no-flow conditions and PO_2 < 6 mm Hg are presented below. The detailed information of sensor construction are given in a submitted paper (4). Fig. 2 shows the structure of the sensor.

Comparison of Lactate Biosensors

Generally, oxidase-based biosensors with amperometric signals generated by either hydrogen peroxide oxidation or oxygen reduction, require molecular oxygen as an electronic mediator. Therefore, the sensor response, for a given substrate concentration, is independent of oxygen tension (PO_2) in the sample, but only, within a certain range. Lower PO_2 values than this range lead to a substantial decrease in the sensor response and its linear dynamic range, as well. A well known strategy to sustain the sensor's linear dynamic range under low PO_2 values is to, deliberately, lower the sensor sensitivity by controlling the diffusion properties of the outer diffusion layer during the sensor construction (5). This is the reason that in our previous effort (3) to monitor lactate under ischemic conditions, the PO_2 values were maintained above 25-30 mm Hg. Although, this range provided an acceptable compromise between sufficient linearity and reasonable sensitivity, these sensors were limited by low signal to noise ratios especially during *in vivo* measurements. Moreover, the effect of a small but significant O_2 tension may have affected the production of lactate. Since anaerobic conditions are the stimulus to produce lactate in ischemic myocardium, it was necessary and important to develop a more relevant lactate biosensor that can function properly under severe oxygen deprivation.

Fabrication, characterization and application of a miniaturized amperometric lactate biosensor are described. The sensor is third generation type, based on crosslinked lactate oxidase and tetrathiafulvalene-tetracyanoquinodimethane (TTF-TCNQ) charge transfer complex. The sensor was developed for continuous quantitative monitoring of lactate accumulation in ischemic myocardium *under severe depletion of oxygen*. The original sensors (3) were made flexible to adhere, or wrap-around the papillary muscle of a rabbit under no-flow, and reduced oxygen environment. This condition of mild ischemia was necessary because the sensors required oxygen as the redox agent or co-factor equivalent to a co-enzyme. Thus the response of the sensor to ambient oxygen was carefully measured. Some calibrations of lactate at decreasing oxygen tensions are illustrated in Fig. 3. Clearly, true ischemia cannot be studied without sensors that operate in the absence of oxygen. Thus the third generation electrodes was constructed, tested, and shown to operate at the lowest oxygen level we could achieve in the experimental chamber containing the perfused, beating rabbit papillary muscle preparation.

Second and third generations of amperometric biosensors overcome the major problem of oxygen limitations associated with the first generation type. Second generation based on non physiological organic molecules, most commonly ferrocene derivatives and tetrathiafulvalene confined to the sensing layer, which acted as redox mediators and replaced the natural oxygen receptor. The major drawback of the mediated biosensor is the leaching of the redox mediator from the vicinity of the electrode surface. Several approaches were introduced to minimize this effect, such as employing electropolymeric film, electrostatic immobilization in an anionic polymer,

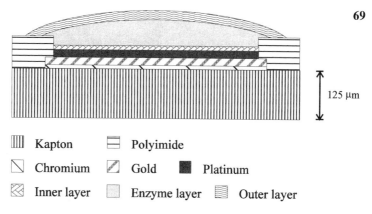

IIII	Kapton	☰	Polyimide		
◻	Chromium	▨	Gold	■	Platinum
▧	Inner layer	▨	Enzyme layer	▤	Outer layer

(1) Layer structure of the first generation type lactate biosensor.

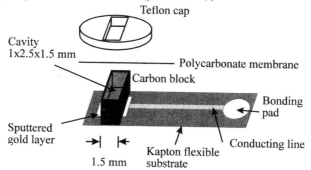

(2) Basic structure of the TTF-TCNQ based lactate biosensor.

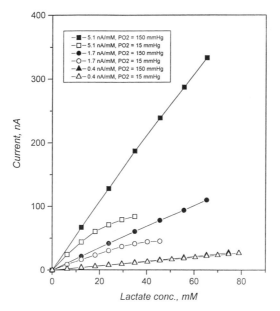

(3) Effect of the sensor sensitivity, in the linear range, on oxygen tension.

crosslinking the redox mediator, attaching of the redox mediator to a polymer, and modification the enzyme itself with mediator molecules. All these modifications impose more complicated steps to the sensor fabrication.

Third generation of amperometric biosensors utilizes organic charge transfer complexes (OCTCs or promoters), such as tetrathiafulvalene– p-tetracyanoquinodimethane (TTF-TCNQ), as an electrode material, instead of conventional Platinum or carbon electrodes. In this case, a direct electron transfer from the enzyme prosthetic group to the electrode surface is achieved without need for redox mediators. OCTCs have been coupled successfully with the three main classes of redox enzymes, i.e., FAD-dependent oxidases, PQQ-dependent dehydrogenases, and NAD-dependent dehydrogenases and with a reductase as well.

A literature search shows there are no reliable reports available till now for developing a robust lactate biosensor based on the third generation type. This is the reason that the objective of this paper is to exploit the unique properties of OCTC, taking TTF-TCNQ as an example, to develop a stable, reliable, lactate biosensor of wide linear dynamic range (~ 20-30 mM), reasonably fast response time and with geometry suitable for the experimental setup used for monitoring extracellular lactate accumulation in ischemic rabbit papillary muscle under no-flow ischemia and severe lowering of oxygen tension. The success of the developed sensors in overcoming the limitations encountered in the previous work is discussed. The content emphasizes the development and application of the TTF-TCNQ-based lactate sensors.

Experimental Section - Sensor Fabrication

Materials, Reagents and Apparatus are comprehensively covered in (*3,4*).
Sensor Fabrication. A cavity of dimensions 1.0 mm (width) x 2.5 mm (length) x 1.5 mm (depth) was drilled in a carbon block of dimensions 1.5 mm x 3.0 mm x 3.0 mm cut from ultrapure carbon rods, 6 mm diameter (Ultra Carbon Corp., Bay City, MI). The carbon block was glued to a sputtered gold electrodes on a flexible polyimide Kapton® foil using silver loaded epoxy (Epoxy Technology, Billerica, MA), see Fig. 2.

TTF-TCNQ crystals were well packed into the cavity using a small stainless spatula, The surface was smoothed against a weighing paper. Three aliquots, 1.5 μl each, of freshly prepared enzyme solution of composition: 6% LOx-3% BSA-0.3% GA were added on to the electrode surface. Each aliquot was allowed to dry for 10 minutes before depositing the next one. A polycarbonate membrane (0.015 μm pore size, Nucleopore, Corning Costar, Livermore, CA) was used as an outer diffusion layer and to protect the contents of the cavity, as well. A Teflon cap (2 mm thick) was used to fix the polycarbonate membrane. The cap was recessed ~ 0.5 mm to allow good contact between the sensor tip and the muscle. Silicone rubber coating (Dow Corning, Midland, MI) was used to isolate the electrical connections. The basic sensor construction is shown in Fig.2.

Lactate Measurement in Control Serum. The human based control sera level II (Dade Moni-Trol• ES Chemistry Control, Baxter Diagnostics, Inc.) were constituted from the lyophilized preparation according to the manufacturer's recommendation. The manufacturer's mean value of lactate concentration is 4.9 mM. A 3 ml PB aliquot was placed in the cell and bubbled with nitrogen for 10 minutes. After background current stabilization, a 1.0 ml aliquot of the serum sample without any treatment, was pipetted into the cell. When the current reached a steady value, the standard addition procedure was used to evaluate the lactate concentration level. The obtained results were compared with the results obtained with the established YSI 2700 SELECT Biochemistry analyzer equipped with YSI 2329 lactate membrane.

Perfusion of Isolated Papillary Muscle. New Zealand white rabbits (n = 3) of either gender weighing 2 to 3 kg were heparinized (200 U/kg, iv) and anesthetized with sodium thiopental (50 mg/kg, iv) in accordance with accepted guidelines for the care and treatment of experimental animals at the University of North Carolina at Chapel Hill. The heart was excised in Tyrode's solution (in mmol/L: Na^+, 149; K^+, 4.5; Mg^{2+}, 0.49; Ca^{2+}, 1.8; Cl^-, 133; HCO_3^-, 25; HPO_4^{2-}, 0.4; glucose, 20) for 20 seconds. The atria and left ventricle (LV) free wall were removed. The LV septal surface of the tissue was pinned to a wax plate in contact with Ag/AgCl ground electrode. The septal artery was visualized with a dissection microscope, cannulated and perfused with a perfusate composed of Tyrode's solution plus insulin (1U/L), heparin (400 U/L), albumin (2 g/L), and dextran (M_r 70,000; 40 g/L). The time elapsed between excising the heart and restarting perfusion was < 5 minutes in each experiment.

The cannula was fixed to the septal artery with a purse string suture and the septum was secured in a custom made experimental chamber. The non-perfused portion of the right ventricle (RV) was excised. One of the underlying papillary muscles was attached by its tendon to a piezosensitve element with a fine silk suture. The perfusate was pumped by a peristaltic pump (Digi-Staltic, Masterflex, Barrington, IL) through a custom made membrane gas exchanger in which partial pressures of O_2, N_2, and CO_2 of the perfusate were controlled. The chamber was closed and the preparation was surrounded by humidified gas with the same composition as the perfusate. The pH of the perfusate was continuously monitored by a pH glass electrode positioned in the perfusion line before entering the septal artery. The amount of CO_2 was adjusted in the gas exchanger to yield a pH of 7.35 ± 0.05. The temperature of the papillary muscle was maintained between 36 and 37.5 °C by passing the perfusate through a thermostated water bath before entering the cannula to the septal artery. The P_{O_2}, P_{CO_2}, HCO_3^- concentration and pH of the perfusate entering the cannula were confirmed by blood gas analysis (System 1304 pH/Blood Gas Analyzer, Instrumentation Laboratory, Lexington, MA). Intraarterial pressure was measured with a pressure transducer (Millar, Houston, TX) and continuously monitored on a strip chart recorder. The septal artery pressure ranged between 35 and 50 mmHg and was maintained by adjustment of the perfusion flow rate (approximately 1.0 to 1.5 ml min^{-1} g^{-1} tissue).

Measurement of Extracellular Lactate. The sensor was positioned in direct contact with the suspended muscle as shown in Fig. 4. A separate platinum wire, serving as a counter electrode, and the miniature Ag/AgCl reference electrode were positioned at the base of the papillary muscle on the septal surface. A 0.7 Hz-5 pole Butterworth low pass filter was used to eliminate the noise resulting from the electrical pulses used for the muscle stimulation. While the muscle was perfused the sensor was polarized at 0.15 V and the background current was allowed to reach a stable value. Ischemia was induced by arresting flow and decreasing the oxygen tension inside the chamber to 3-6 mm Hg. The volume fractions of O_2 and CO_2 in the recording chamber were measured using the gas analyzer. The magnitude of lactate accumulation was monitored by recording the change of the sensor output against the time after arrest of flow.

To convert the current response into concentration, the sensor was calibrated *in situ*, at the end of the experiment, by perfusing the muscle with 7.5 and 15 mM lactate standards prepared in the perfusate solution. The sensor response to these standards under flow conditions was used to construct a two-point *in situ* calibration curve. Ideally, the sensor should be calibrated under the same experimental conditions, i.e., no-flow conditions. However, such a calibration will not be reliable because under no-flow conditions the muscle will generate lactate. In this way the standard value will be completely uncertain. For this reason the sensor calibration was performed under flow conditions. We believe that such *in situ* calibration is more accurate than the *in vitro* calibration approach which is accepted when a calibration under identical condition is not possible, e.g., with *in vivo* measurements.

Results and Discussion

Sensor Design. In terms of geometry, the planar sensors used in the previous work (*3*) offered an advantage of having the sensing part on one side of the sensor rather than on the bottom as in the conventional electrodes design. This is of especial importance in positioning the sensor through a small window in the top of the experimental chamber, and allowing simultaneous top view monitoring of the preparation along with the sensors using a camera mounted above the chamber window. To keep the desirable configuration and flexibility, the sputtered gold electrodes designed for the previous work (*3*) realized on flexible Kapton substrate were adapted in the present work.

Simple deposition of TTF-TCNQ on the gold electrode surface using the drop coating technique is not feasible because of the poor adhesion to the electrode surface. More important, use of outer diffusion layer, e.g., polycarbonate membranes on this planar thin substrate becomes impossible. To promote a more convenient geometry for the base electrode, a carbon block with self contained cavity was glued to the electrode surface using silver epoxy. The dimensions of the sensing tip (1.5 mm x 3.0 mm) was designed to fit the dimensions of the rabbit papillary muscles, about 1.2 mm in diameter and 4.0 mm length, studied in the present work as shown in Fig. 4.

Sensor Fabrication and Characteristics. In most of the previous work, conducting salts were mixed with a polymer binder such as poly (vinyl chloride) or more commonly, an oil to enhance the consistency of the electrode surface This treatment is not appropriate in the present case because the sensor will be in direct contact with a beating muscle. An outer protecting membrane that provides a diffusion barrier as well for lactate is essential for the sensor construction. Microporous polycarbonate (PC) membranes were used successfully for these purposes. The PC membrane, acting as a diffusion barrier, enhances the linear dynamic range by lowering the lactate flux that reaches the enzyme layer.

Sensor Stability. The sensors maintained almost 90% of the initial activity after three weeks, stored at 4 °C when not in use. The short term operational stability of the sensor was assessed by polarizing the sensor at 37 °C in a stirred PB solution for more than 6 hours. No observed change in sensitivity was observed after this treatment. Both storage and operational stability of the sensor are excellent for the present purpose since the entire experiment time is usually in the range of 2 hours The enhanced stability is believed to be due to the combined immobilization, i.e., adsorption and glutaraldehyde crosslinking with BSA.

Effect of Applied Potential on the Sensor Response. Generally, TTF-TCNQ electrodes show a potential stability window from +0.5 to -0.2 V *vs*. SCE (*6*). Therefore, the applied potential must not exceed these limits. The useful operating potential ranges used for sensors were limited to a narrower range, to achieve a compromise between sensor sensitivity, selectivity, and stability (*7*). The effect of the applied potential on the sensor response is shown in Fig. 5. Generally, the linear range, sensitivity, and background current increased with increasing the applied potential. Applied potential in the range of 0.1 to 0.15 V offer a good compromise between low background current, high linear dynamic range, selectivity and good stability. Therefore, an applied potential of 0.15 V was used in all subsequent measurements. A similar operating potential range was reported for TTF- TCNQ based glucose sensor (*8,9*).

Effect of Oxygen and Temperature. In principle, second and third generations of amperometric biosensors do not rely on oxygen as electron carrier. However, the presence of oxygen produces an adverse effect on the current response as shown in Fig.6 (n=3, error bars were omitted for clarity). Curve A shows a typical attenuated nonlinear response for low substrate concentrations observed in the presence of oxygen using such non-oxygen mediated biosensors (*9,10*). This effect results from the competition between oxygen and oxidized mediator (second generation) or the electrode reaction (third generation) to regenerate the reduced enzyme. At higher concentrations, Curve A (air-saturated) and B (nitrogen-saturated) approach each other because of oxygen depletion in the immediate environment of the enzyme layer. Under the conditions of curve B, the sensors showed a linear dynamic range up to 22 mM with sensitivity of 166 ± 39 nA/mM. Curve C represents the sensor calibration at 37 °C in nitrogen-saturated PB solution. The sensors showed higher sensitivities (380 ± 94 nA/mM) and enhanced linear dynamic range up to 27 mM.

Apparently, the conditions of very low or absence of oxygen and the physiological temperature (37 °C) are favorable for the sensor response in terms of higher sensitivity and best linearity. Therefore, the present application of monitoring lactate under ischemic conditions can be considered as the most relevant application, till now, that uses the maximum advantage offered by non-oxygen mediated biosensors where the oxygen during ischemia is controlled to a very low and constant value. The problem of oxygen depletion or fluctuation has not been totally solved by non-oxygen mediated biosensors in other *in vivo* applications.

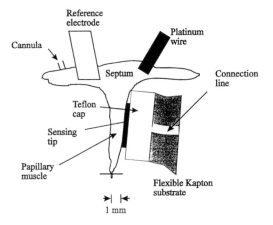

(4) Sketch of a top view of an isolated perfused RV rabbit papillary muscle suspended in an artificial atmosphere and in contact with rectangular sensing tip of the lactate sensor. A miniature Ag/AgCl reference electrode is fixed on the muscle base. A 1 mm platinum wire (serving as counter electrode) is fixed on the septum. Electrical stimulation is delivered at twice diastolic threshold through a Pt wire affixed to the tendon.

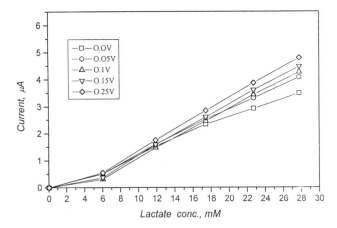

(5) Effect of applied potential on the sensor response.

Interferences. Interference by easily oxidizable species such as ascorbic acid, uric acid, L-cysteine, acetaminophen is not probable in the present application because the perfusate solution does not contain any of them. However, the sensor response to the most common electroactive species that usually exist in the biological samples was measured for possible future applications. The results show that only ascorbic acid gives a significant effect. This ascorbate interference can be attenuated to a great extent by a well known strategy using an additional layer of ascorbate oxidase (11).

Effect of pH. The sensor response to 15 mM lactate in phosphate buffers of different pH values was measured. The sensor output (n = 3) is constant to 5 ± 1.5 % in the pH range 6.0 - 8.5. This range covers the physiological region of interest.

Lactate Measurement in Control Serum. The reliability of the sensor for lactate monitoring in a biological complex matrix was assessed by determining lactic acid in a human based control serum which contained 42 constituents. The results obtained with the present sensor at 37 °C and nitrogen bubbling were compared with those obtained with the YSI Biochemistry Analyzer. The excellent agreement (average recovery = 103.5 ± 0.8 %) demonstrated that our sensor is reliable for lactate determination in real samples under conditions similar to those used in the papillary muscle experiments.

Time Course and Magnitude of the accumulation of Extracellular Lactate in Rabbit Ischemic Ventricular Muscle. The performance characteristics of the developed sensor, such as operational stability, high linearity, and ideal response under low oxygen tensions suggest its successful use in continuous monitoring of the extracellular lactate accumulation in an ischemic rabbit papillary muscle. According to the authors' experience, this is the first attempt on continuous lactate monitoring using an amperometric sensor located in direct contact with the myocardium.

The lactate content of the perfusate was measured and found to be zero prior to the onset of no-flow ischemia. As shown in Fig. 7 the accumulation of extracellular lactate reached 16.4 ± 2.9 mM. after 15 minutes of no-flow ischemia. The perturbation observed on arresting the flow and reperfusion on the previous work, is no longer encountered using the TTF-TCNQ sensor.

Conclusion

Both of the new lactate biosensors show many attractive features, such as miniaturization, flexibility, wide linear dynamic response range at low oxygen tensions, good stability, adequate sensitivity, fast response time and selectivity. Their special geometries are important in monitoring successfully extracellular lactate accumulation during no-flow ischemia. These novel lactate sensors will be a useful tool for further detailed studies in which K^+, pH, and lactate are monitored simultaneously for longer periods of ischemia, and to study lactate accumulation in hypoxic conditions, as well as conditions that simulate the boundary between normal and ischemic tissue. These conditions are characterized by low oxygen tensions and accumulation of potassium.

Acknowledgments. S. A. M. Marzouk gratefully acknowledges Ain-Shams University, Cairo, Egypt for providing resources for his stay at University of North

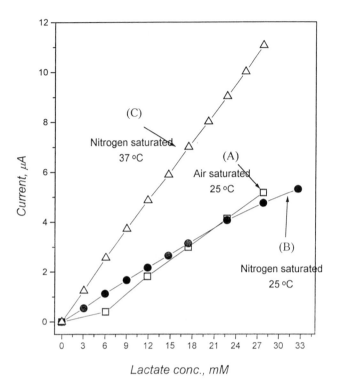

(6) Effect of oxygen and temperature on the sensor response.

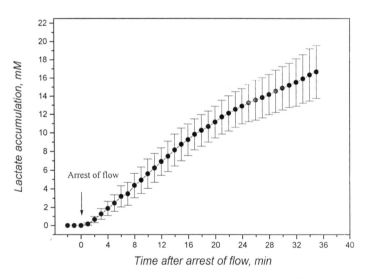

(7) Extracellular lactate accumulation during 35-minute of no-flow ischemia. PO2 values were controlled to < 6 mm Hg. Data obtained from 3 different animals and three different sensors.

Carolina. Authors like to acknowledge R. B. Ash for the wafer fabrication and D. Sandiford for technical assistance. This work was supported, in part, by grants RO1 HL48769 and P01 HL27430 from the NIH, HLBI, Bethesda, MD.

Literature Cited.

(1) Neely, J. R.; Whitmer, J. T.; Rovetto, M. J. *Circ. Res.* **1975**, *37*, 733.
(2) Kabayashi, K.; Neely, J.R. *Circ. Res.* **1979**, *44*, 166.
(3) Marzouk, S. A. M.; Cosofret, V. V.; Buck, R. P., Yang, H.; Cascio, W. E.; Hassan S. M. *Talanta* **1997**, in press.
(4) Marzouk, S. A. M.; Cosofret, V. V.; Buck, R. P., Yang, H.; Cascio, W. E.; Hassan S. M. Anal. Chem., **1996**, submitted.
(5) Hu, Y.; Zhang, Y.; Wilson, G. S. *Anal. Chim. Acta* **1993**, *281*, 503.
(6) Jaeger, C. D.; Bard, A. J. *J. Am. Chem. Soc.* **1979**, *101*, 1690.
(7) Khan, G. F. *Biosensors & Bioelectronics* **1996**, *11*, 1221.
(8) Khan, G. F.; Ohwa, M.; Wernet, W. *Anal. Chem.* **1996**, *68*, 2939.
(9) Kawagoe, J.; Niehaus, D. E.; Wightman, R. M. *Anal. Chem.* **1991**, *63*, 2961.
(10) Palleschi, G.; Turner, A. P. F. *Anal. Chim. Acta* **1990**, *234*, 459.
(11) Nagy, G.; Rice, M. E.; Adams, R. N. *Life Science*, **1982**, *31*, 2611.

Chapter 7

Neural Networks and Self-Referent Acoustic-Wave Sensor Signaling

Lan Bui and Michael Thompson

Department of Chemistry, University of Toronto, 80 St. George Street, Toronto, Ontario M5S 3H6, Canada

This paper presents the application of artificial neural networks (ANN) to the self-referent calibration of thickness-shear mode (TSM) acoustic wave chemical sensors. Spectrum analysis of impedance measurements affords complete characterization of the TSM sensor, which includes the use of the Butterworth - Van Dyke (BVD) equivalent circuit to quantify the electrical responses. The multidimensional nature of this method and a novel weight-adjustment procedure, applied to ANN calculations, are utilised to effect a method of calibration in the presence of interferents. A network is trained, using exemplary I/O data acquired for a potassium chloride (KCl) system, to predict unknown outputs ie. concentration, given four sets of measured inputs ie. series resonant frequency (Fs), parallel resonant frequency (Fp), motional capacitance (Cm) and motional resistance (Rm). The trained and tested network achieved a high predictive efficiency with errors in the range of 2%-6%. An interferent, ethanethiol, is added to test the robustness of the trained network and was found to adversely affect the predictive ability of the network. The magnitudes of the weights, which are associated with the set of inputs deemed to be most affected by the interferent (Fs) are adjusted to minimise this deterioration. The resultant network, calibrated for the interferent, achieved the same predictive efficiency for adulterated samples as that achieved by the original network for unadulterated samples.

A major problem since the inception of chemical and biosensor technology has been the quality of the signal-concentration relationship with respect to the time course of

©1998 American Chemical Society

measurements. Lack of reproducibility in concentration values can be traced to changes in the performance characteristics of the device concerned over time and possible perturbations caused by matrix interferences. The incorporation of an artificial neural network (ANN's) into a sensor system may assist in solving this problem by providing a self-referent calibration mechanism for matrix effects. The main advantage of the neural network calculation over classical regression analyses is its ability to model at an implicit level any number of complex, non-linear data sets that would be impossible, or at least extremely difficult, to fit to an explicit mathematical model. This advantage has resulted in the application of neural network analyses to numerous fields among them, spectral interpretation *(1-4)*, process control *(5-7)*, prediction of protein structure *(8-9)* and various sensor systems *(10-17)*. ANN's have also been utilised in piezoelectric-based sensor systems but the focus has been primarily on arrays of single-output sensors, with the output usually being the series resonance frequency (Fs) *(18-22)*. This single-output array architecture, in addition to generating extra complexity in theory and application, has served to limit the utility of neural networks in pattern recognition and classification. In the present work, spectrum analysis of the impedance measurements of a single sensor, using a network analyser, provides multidimensional information thus effectively overcoming these drawbacks. Such multi-dimensionality, in conjunction with a novel weight-adjusment procedure, may provide a calibration mechanism that can be used to eliminate a number of matrix effects.

Network Analysis To Provide Multi-dimensionality. A typical response pattern derived from network analysis of a TSM device, operated in liquid, is depicted in Figure 1. The theory behind network analysis of the TSM sensor has been discussed thoroughly by various investigators and, thus, will only be briefly summarised here *(23-28)*. Several different parameters can be obtained from the plots of phase angle and impedance versus frequency. These can be classified into two groups: frequency-based parameters such as the series resonant frequency and the parallel resonant frequency, and impedance-based quantities such as the maximum impedance (Zmax), the minimum impedance (Zmin) and the maximum phase angle (θmax). Approximations in analysis of the electrical behaviour of the device, using the Butterworth-Van Dyke (BVD) equivalent circuit (Figure 1), can provide four more useful parameters: the static capacitance (Co) between the two parallel electrodes on the two surfaces of the quartz crystal as well as the motional resistance (Rm), the motional capacitance (Cm) and the motional inductance (Lm), all of which arise from electromechanical perturbation induced by the piezoelectric effect in the resonant region.

Since neural calculations are inherently complex and data-intensive, this study is purposely focussed on a relatively simple system; the responses of the TSM device to an electrolyte, potassium chloride (KCl). For the same reason, only four of the more important parameters are selected to be used in the initial phase of this project. The four chosen quantities are the series resonant frequency (Fs) which represents mass loading and other interfacial effects, the parallel resonant frequency (Fp) which characterises the capacitive loading effect, the motional resistance (Rm) which is related to the energy dissipation of the oscillating sensor and the static capacitance (Co) which reflects

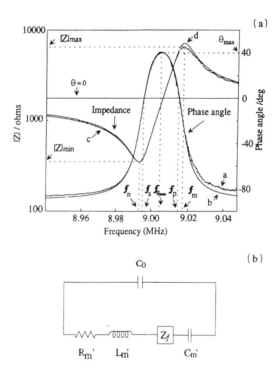

Figure 1. (a) Network analysis traces and (b) equivalent circuit parameters.

changes in the interfacial capacitance. Exemplary I/O data, comprising the chosen parameters is used to train a network, followed by appropriate testing. Once the optimal architecture is found, the trained and tested network is used to predict concentrations of various "unknown" solutions, given the respective inputs. In the second phase of the project, an interferent, hypothesised to affect only one of the parameters, is added to the "unknown" samples in order to test the robustness of the trained network. The adulterant of choice was ethanethiol (CH_3CH_2SH), which is postulated to affect mainly the series resonant frequency (Fs) through its absorption to the surface of the gold electrode through the well-known Au-S interaction. It has been shown that Rm and Co are not affected by a change in surface mass since the deposited layer can be regarded as an extension of the material comprising the electrode (24). Thus, power dissipation, reflected by Rm, and the dielectric properties of the quartz plate, reflected by Co, should remain constant. The parallel resonant frequency, Fp, is also affected by mass deposition at the surface but it is not as sensitive to mass changes as Fs (29). Adjustments of the weights associated with affected nodes are then performed to calibrate for the effect of the adulterant. An analogous calibration scheme may also be possible for an array of single-output sensors, but this methodology is expected to be far more complex with respect to theory and application.

Experimental

AT-Cut quartz piezoelectric crystals coated with gold electrodes were obtained from Classic Frequency Control, Inc., Oklahoma City, OK, USA. The resonant frequencies of the crystals were ~ 9 MHZ and the electrode surfaces were polished to < 1μm. The TSM device was clamped in a cell with O-rings on both sides. One side of the crystal was kept under nitrogen while the other side was immersed in the analyte solution. Potassium chloride and ethanethiol were used as received from Aldrich. Doubly distilled, deionised water was used as the reference and as the solvent for the standard solution. Network analysis was performed using a Hewlett Packard HP 4195A network/spectrum analyzer, equipped with an HP 41951A impedance test kit and an HP 16092A spring clip fixture. All data were transferred to an IBM-compatible computer through an IEEE-488 bus.

Two different neural network packages were utilised in this project. The first, ANNAT, originates from collaborators in Åbo Akademi, Finland. This program uses the Levenberg-Marquardt algorithm for optimization which makes use of an interpolation between the steepest descent and the Newtonian directions, based on a "trust region" where a linearised model is assumed to apply. The software is simple-to-use, fast and robust but it does possess some drawbacks, mainly that the number of weights (connections) should be kept below 100 and that the architecture of a network must have the same number of nodes in each hidden layer. The second program is a commercial package, NeuralWorks Professional II/Plus v. 5.22, from NeuralWare, Pittsburgh, PA, USA. Several algorithms for RMS optimization are offered but only back-propagation was utilised. This package does not possess the limitations of the in-house program but the typical back-propagation disadvantages apply here, that is, longer training and testing times and a tendency towards local minima.

Neuralnet Results for Unadulterated Systems. Training and testing data sets are constituted of 100 and 50 instances, respectively, covering the range of concentrations between 10^{-4}M and 1M. The concentrations of the "unknown" samples are also in this range. A typical network, with a 4,6,1 architecture: 4 inputs, 6 nodes in one hidden layer and 1 output, is shown in Figure 2. A bias is used to ensure the nodal outputs are in the optimal range and the window in the top left hand corner tracks the decreasing RMS error. It can be seen that for a typical viable network, the RMS decreases in a near exponential fashion.

A typical output of a trained ANNAT network is depicted in Table I. This represents a 4,5,1 network with 30 weights reaching convergence in 11000+ iterations. Each of the weights can be related back to a specific connection in the network. The relevance of this is that the weights can be traced back to a specific input and this will be important for the weight adjustment process. The sum of squares error (SSQ) shown was converted to RMS for meaningful comparison. The optimal architecture of any network is usually derived through a trial and error process. Figure 3 illustrates the relationship between RMS and the number of hidden nodes for a network with one hidden layer. The RMS shows an initial rapid improvement with increasing number of hidden nodes, followed by a slower and decreasing phase of improvement. After a certain number, the slight improvements no longer justify the extra computing time and effort, and the network is set. If testing procedures prove the network to be unsuitable then a different architecture may be chosen.

Using the ANNAT package, the best architecture obtained is a 4,7,1 net with no further improvement after 10000+ iterations. The RMS for the test set can be seen to be of the same magnitude as that of the training set, validating the network (Table II). For the commercial package from NeuralWare, the best architecture is 4,9,1 and the number of iterations required is approximately doubled. The RMS for the training and testing sets are also of similar magnitudes, indicating a useful architecture. Slightly lower training RMS was observed for NeuralWorks but the test RMS for this program was actually slightly higher than that of ANNAT. The discrepancy is not large but it could be argued that the larger NeuralWorks network, although it may have trained better, may not generalise as well as the smaller ANNAT architecture.

Twenty solutions of varied concentrations in the 10^{-4} M to 1 M range were chosen to be the unadulterated "unknown" set. The response sets for these "unknowns", comprising Fs, Fp, Co and Rm, were used to test the predictive ability of the trained networks. Using ANNAT, these predictions resulted in errors in the range of 3-6%, compared to the true values (Table III). This can be considered to be acceptable given the large range of concentrations being studied. A network optimised for a smaller range of concentrations will undoubtedly yield a corresponding reduction in the error range. The predictive ability of the commercial package, with respect to the unadulterated "unknown" data sets, is slightly better than that of ANNAT (Table IV). This is somewhat surprising in light of the higher test RMS for NeuralWorks. The

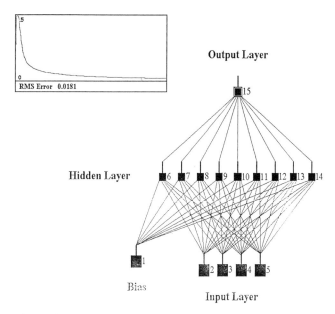

Figure 2. Architecture of a neural network extracted from NeuralWorks.

Table I. Typical Output for Annat Training Run

Number of input and output nodes:	4	1	
Number of hidden layers:	1		
Number of nodes in each hidden layer:	5		
Activation function for output nodes:	1		
Activation function for hidden nodes:	1		
Gain term - beta:	1.0000		
File with input patterns and outputs:	kcl.dat		
Print option (0/1/2):	0		
Maximum number of iterations:	20000		
Analytical/numerical derivatives <A>:	A		
Limit for initial weight-guesses:	.00000E+00		
Seed:	9823		

Number of patterns in the input file: 100

The weights W >>> 1 to 30 :

7.6818	-.31707E-01	.69635E-02	-.14397	3.6978
-48.077	-.12729E-01	-.42298E-02	.23473	7.0486
7.304	.22442E-01	-.13044E-02	.86784E-01	-3.7655
1096.1	-1097.0	-1.0025	-4.8776	1.0893
-25.790	.90235E-01	-6.0004	-.34710E-0	291.772
3.1639	-66.097	.36891E-01	.10492	-69.200

SSQ :   .44755E-03

Number of iterations : 11124

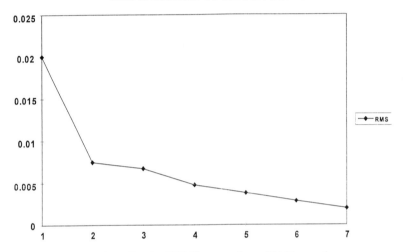

Figure 3.   Graph of RMS vs number of hidden nodes.

Table II. Comparison of Software Packages

Parameter	ANNAT	NeuralWorks
Best architecture	4,7,1	4,9,1
Number of iterations	10000+	20000+
RMS (train)	0.00142	0.00129
RMS (test)	0.00407	0.00445
Predictive error (unadulterated)	3%-6%	2%-5%
Predictive error (thiol, unadjusted)	8%-12%	7%-14%
Optimal weight reduction	18%	21%
Predictive error (thiol, adjusted)	3%-7%	3%-6%

- All errors are absolute.

Table III. Predictive Results for ANNAT

Molarity	% Error (unadulterated)	% Error (adulterated)	
(M Kcl)		non-adjusted	weight-adjusted
0.0005	6.2	11.5	6.7
0.005	5.9	11.0	7.0
0.015	5.1	11.9	7.4
0.025	4.7	10.6	6.0
0.035	3.6	9.4	5.2
0.045	4.5	9.7	5.1
0.055	3.7	9.1	4.5
0.065	3.5	9.2	5.7
0.075	3.4	8.4	4.0
0.085	3.6	7.9	3.5
0.095	3.3	8.7	3.1
0.150	3.7	8.1	4.2
0.250	3.0	8.9	2.9
0.350	4.2	9.5	3.5
0.450	4.7	8.6	4.0
0.550	5.5	10.0	3.7
0.650	3.9	9.2	4.5
0.750	4.9	8.8	3.9
0.850	4.3	9.5	4.5
0.950	3.6	9.0	3.5

- All errors are absolute.

Table IV. Predictive Results for NeuralWorks

Molarity	% Error (unadulterated)	% Error (adulterated)	
(M KCl)		non-adjusted	weight-adjusted
0.0005	5.4	13.5	5.8
0.005	4.1	13.9	6.0
0.015	4.8	12.7	5.4
0.025	3.8	11.6	5.1
0.035	4.3	10.2	5.2
0.045	3.5	9.9	4.1
0.055	3.9	9.9	3.5
0.065	3.0	9.2	2.7
0.075	2.4	7.6	4.0
0.085	3.6	6.9	3.9
0.095	3.1	8.7	3.7
0.150	1.9	7.1	4.2
0.250	3.8	9.9	4.9
0.350	3.0	9.1	3.8
0.450	4.2	9.6	4.0
0.550	3.5	11.0	4.7
0.650	2.9	7.7	4.6
0.750	4.0	9.8	4.9
0.850	4.3	9.2	4.2
0.950	3.6	10.0	3.9

- All errors are absolute.

differences are small enough that they can be attributed to normal distribution and/or experimental errors. For both systems, no clear trend in the % error can be observed, with respect to concentration, but the highest % errors seem to occur for lower concentrations.

Neuralnet Results for an Adulterated System. Neural network analysis has been shown to be effective in handling ideal data sets but this alone does not justify the time and effort spent creating the appropriate architectures. The next step involves the exposure of the trained networks to data generated by samples that have been adulterated with an interferent, ethanethiol (CH_3CH_2SH). This compound was chosen since the thiol functional group will interact with the gold electrodes of the TSM device to effect a change in the series resonant frequency. The thiol was introduced into samples of the same KCl concentrations as those in the unadulterated "unknown" set. Although a larger molecule will cause a larger deviation in Fs, ethanethiol was chosen in view of its solubility in an aqueous system. The shorter chain length also precludes formation of a well-packed monolayer. This, combined with a purposely low concentration of thiol (<mmol), ensures that changes in the viscosity, density and electrical conductivity of the sample are kept to a minimum so that the change in Fs can be attributed mainly to mass loading. A secondary effect is that the other three parameters, Rm, Co and Fp will remain relatively unperturbed.

As can be seen from Table II, the predictive errors for the adulterated "unknown" set are significantly higher than those for the unadulterated "unknown" set, between 8-12% for ANNAT and 7-14% for NeuralWorks. These results established the neural networks is fairly robust, but it is obvious that a network trained using unadulterated I/O data cannot be used to predict adulterated samples accurately. Once again, no clear trend for the % error was observed, with respect to concentrations, but the predictive errors for NeuralWorks were found to be slightly higher than those for ANNAT. This seems to indicate that while the larger network may generalise an ideal data set better, it does not possess the same robustness for tainted data.

Weight Adjustment. As predicted earlier, the changes in Fs were found to be significantly larger for the adulterated runs, compared to the changes in the other three parameters. It is clear that the degradation of the performance of the trained networks can be attributed mainly to the change in Fs. Several possible procedures can be considered to rectify this situation. The elimination of Fs as an input would be an effective remedy but this would involve repeating the training and testing processes to find the optimal architecture for a three-inputs system. Another solution is to increase the number of inputs, making use of the other available parameters, in order to reduce the effect Fs has on the trained network. This may indeed increase the robustness of the network but once again it involves time-consuming training and testing procedures which are certain to be more intensive due to the extra complexity introduced by the larger data set.

For any trained network, each of the weights can be related back to a specific connection in the neuralnet. It was hypothesised that manipulation of the weights, which are associated with the Fs-to-hidden layer connections, may provide a possible modulation mechanism of the Fs input and this, in turn, may temper the effect of the adulterant. It was discovered that a reduction in the magnitudes of these weights resulted in an improvement in prediction accuracy for the adulterated samples for both software packages. Optimal results were obtained with reductions of approximately 20% in the magnitudes of the affected weights; 18% and 21% for ANNAT and NeuralWorks, respectively (Table II). The similarity in the magnitudes of the adjustment indicates that this is a systematic phenomenon and not a random one. The predictive errors achieved for the adjusted networks are close to those for the unadulterated samples: 3-7% for ANNAT and 3-6% for NeuralWorks, respectively (Table III and IV). The small decrease in efficiency may be attributed to the fact that Fp is also affected by changes in surface mass, albeit to a smaller degree.

The manipulation of the weights associated with the affected parameter seems to have lessened the deleterious effect of the interferent without having to resort to the construction of new networks. In effect, the existing networks have been calibrated for the presence of the thiol. While the results are encouraging, it must be stated that more work needs to be done before this weight-adjustment procedure is better understood and validated. Research in our laboratory is proceeding with regard to the manipulation of the weights associated with the other three parameters. Future experiments will include the introduction of a second and further target interferents.

Conclusions

To summarise, back-propagation networks yield better results but at the cost of longer computing time. The inclusion of more relevant parameters, other than the chosen four, will help to lower the errors and improve robustness, but at greater computing cost. The need to gain a better understanding of the weight adjustment process is extremely important and this can be done by manipulating various sets of weights followed by observation of the resultant changes in predictive accuracy. If this process can be elucidated, it can be used in a manner roughly analogous to the temperature calibration process of a pH meter. A multiple-points calibration is also feasible if the effects of the interferent are not isolated to one parameter and this may be evident in the next step, the introduction of multiple interferents. Once the study on this relatively simple electrolyte system is completed, the final step will be the application of this process to various chemically selective sensors already developed in our laboratory such as genosensor, conductive polymer and self-assembled monolayer-based systems.

Acknowledgment

We are grateful to the Natural Science and Engineering Research Council of Canada for support of this work.

Literature Cited
1. Robb, E. W.; Munk, M.E. *Mikrochim. Acta* **1990**, 131.

2. Anker, L. S.; Jurs, P. C. *Analytical Chemistry* **1992**, 64, 1157.
3. Lin, C. W.; La Manna, J. C.; Takefuji, Y. *Biological Cybernetics* **1992**, 67, 303.
4. Tanabe, K.; Tamura, T.; Uesaka H. *Applied Spectroscopy* **1992**, 46, 807.
5. Hoskins, J. C.; Himmelblau, D. M. *Comput. Chem. Eng.*, **1988**, 12, 881.
6. Watanabe, K.; Matsuura, I.; Abe, M.; Kubota, M.; Himmelblau, D. M. *AIChE J.* **1989**, 35, 1803.
7. Bhat, N.; McAvoy, T. *J. Comput. Chem. Eng.* **1990**, 14, 573.
8. Kneller, D. G.; Cohen, F. E.; Langridge, R. *J. Mol. Biol.* **1990**, 214, 171.
9. Bohr, H.; Bohr, J.; Brunak, S.; Cotterill, R. M. J.; Fredholm, F., Lautrup, B.; Olsen, O.H.; Petersen, S. B. *FEBS Lett.* **1990**, 261, 43.
10. Bos, M.; Bos, A.; Van der Linden, W. E. *Anal. Chim. Acta* **1990**, 233, 31.
11. Walt, D. R.; Dickinson, T. A.; Healy, B. G.; Kauer, J. S.; White, J. *Proc. SPIE-Int. Soc. Opt. Eng.* **1995**, 2508, 111.
12. Ganesh, C.; Steele, J. P. H.; Zhang, H.; Mishra, D.; Jones, J. *Int. SAMPE Symp. Exhib.* **1994**, 39, 883.
13. Fujii, S.; Kamada, T.; Hayashi, S.; Tomita, Y.; Takayama, R.; Hirao, T.; Nakayama, T.; Deguchi, T. *Proc. SPIE-Int. Soc. Opt. Eng.* **1995**, 2332, 612.
14. Nanto, H.; Tsubakino, S.; Kawai, T.; Ikeda, M.; Kitagawa, S.; Habara, M. *J. Mater. Sci.* **1994**, 29(24), 6529.
15. Wang, X.; Yee, S.; Carey, P. *Sens. Actuators B* **1993**, 13(1-3), 458.
16. Tonoike, M. *Cosmet. Toiletries* **1990**, 105(6), 49.
17. Sah, W. J.; Lee, S. C.; Tsai, H. K.; Chen, J. H. *Appl. Phys. Lett.* **1990**, 56(25), 2539.
18. Chang, S. M.; Muramatsu, H.; Karube, I. *Sens. Mater.* **1995**, 7(1), 13.
19. Hivert, B.; Hoummady, M.; Henrioud, J. M.; Hauden, D. *Sens. Actuators, B* **1994**, 19(1-3), 645.
20. Moriizumi, T.; Nakamoto, T.; Sakuraba, Y. *Olfaction Taste XI, Proc. Int. Symp., 11th* **1994**, 694.
21. Nanto, H.; Tsubakino, S.; Ikeda, M.; Endo, F. *Sens. Actuators B* **1995**, B25(1-3), 794.
22. Pisanelli, A.; Qutob, A. A.; Travers, P.; Szyszko, S.; Persaud, K. *Life Chem. Rep.* **1994**, 11(2), 303.
23. Kipling, A. L.; Thompson, M. *Anal. Chem.* **1990**, 62, 1514.
24. Martin, S. J.; Granstaff, V. E.; Frye, G. C. *Anal. Chem.* **1991**, 63, 2272.
25. Yang, M.; Thompson, M. *Anal. Chim. Acta.* **1992**, 269, 167.
26. Yang, M.; Thompson, M. *Anal. Chim. Acta.* **1993**, 282, 505.
27. Martin, S. J.; Frye, G. C.; Ricco, A. J.; Senturia, S. D. *Anal. Chem.* **1993**, 65, 2910.
28. Schneider, T. W.; Martin, S. J. *Anal. Chem.* **1995**, 67, 3324.
29. Yang, M.; Thompson, M. *Anal. Chem.* **1993**, 65, 1158.

GAS SENSORS AND
THEIR APPLICATIONS

Chapter 8

Design and Testing of Hydrogen Sensors for Industrial Applications

B. S. Hoffheins, R. J. Lauf, T. E. McKnight, and R. R. Smith

Oak Ridge National Laboratory, Oak Ridge, TN 37831–6004

Many potential applications for hydrogen detectors demand low-cost, low-power devices that are small, reliable, and rugged. A solid-state sensor developed at Oak Ridge National Laboratory has the potential to meet these requirements. The sensor mechanism relies on the well-characterized solubility of hydrogen in palladium. Changes in hydrogen partial pressure can be measured by corresponding changes in the electrical resistance of the sensor's palladium metallization. The sensor is fabricated using traditional thick-film manufacturing practices, which are inexpensive and, more importantly, have inherent design flexibility. We have recently characterized the sensor performance under a range of conditions, including elevated temperature and humidity. We describe these tests and give future directions as to sensor design, mass production considerations, and feasibility in end-use applications.

The scientific basis for many hydrogen sensors relies on the reversible solubility of hydrogen in palladium (1-7). Most hydrogen partial pressures of interest (10^{-3} to 1 psia) will be in equilibrium with the palladium-hydrogen solid solution (α phase) at ambient temperatures. In this range, hydrogen molecules dissociate on the Pd metal surface and atomic hydrogen is quickly absorbed as an interstitial solute in the Pd until equilibrium is reached. The dissolved hydrogen increases the electrical resistivity of the Pd in a known, quantitative way (8). A thin-film hydrogen sensor

* Work supported by Electric Power Research Institute, Inc. and U.S. Department of Energy, Office of Energy Research, Science and Technology Partnerships.
† Managed by Lockheed Martin Energy Research Corporation for the U.S. Department of Energy under Contract No. DE-AC05-96OR22464.

©1998 American Chemical Society

based on this phenomenon was developed at ORNL and demonstrated for hydrogen concentrations up to the lower explosive limit (LEL) in air (about 4%) (9).

After some preliminary testing of the thin film sensor, we recognized that a thick-film sensor based upon the design of the original thin-film device would be inherently simpler, more robust, and cheaper to manufacture. Most of the needed thick-film compositions, or "pastes," were off-the shelf products; however, the sensor required one completely new paste containing pure Pd and a very durable glass frit. We successfully formulated a sensor paste by combining a commercial Ag/Pd-based conductor composition with fine Pd power and then teamed with DuPont Electronics (Research Triangle Park, North Carolina) to specify and develop an optimized composition for mass production (10-11). In preliminary tests, these thick film sensors had a linear response to increasing concentrations of hydrogen (0.5% to 4% in argon/air matrix).

The sensor design is similar to that of a classical strain gauge with active and reference components. Four palladium conductors are deposited as serpentine patterns and connected as a Wheatstone bridge to increase measurement sensitivity and build in temperature compensation, Figure 1. One or two of the palladium legs are "active" and are exposed to ambient conditions. The remaining serpentine legs are "reference" legs and are passivated by a dielectric layer impermeable to hydrogen. Dc power is applied to opposite nodes of the circuit and the bridge voltage is measured across the remaining nodes, as indicated in Figure 1.

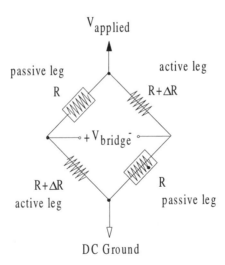

Figure 1. Schematic representation of the hydrogen sensor.

In a typical bridge circuit, the objective is to measure the resistance change of one resistor. The model for this change is:

$$V_{bridge} = \frac{V_{applied}}{2}\left(\frac{\Delta R}{2R + \Delta R}\right),$$

where R + ΔR is the resistance of the active leg in the presence of hydrogen and V_{bridge} and $V_{applied}$ correspond to sensor voltages across the bridge, as indicated in Figure 1. For a sensor circuit, the sensitivity can be doubled by having two active resistor legs. In this case, the equation becomes:

$$V_{bridge} = V_{applied}\left(\frac{\Delta R}{2R + \Delta R}\right).$$

The prototype thick-film sensor design consists of four distinct patterns, separately screen-printed and fired on a 645-mm^2 (1-in.^{2}) alumina substrate (Figure 2). The application of each pattern proceeds in order of firing temperature, from highest to lowest. The palladium sensor metallization (DuPont DCH012) is fired first at 1000°C. The serpentine path length in current prototypes is 20 cm and measures between 70 and 100 ohms. A high resistance path not only facilitates sensitivity but also decreases power consumption, making continuous battery operation feasible. (Fortunately, we have not yet approached thick film line width limitations and are designing a new layout for even greater sensitivity and lower power requirements.) The second printing (DuPont 6120), fired at 900-950°C, forms silver-palladium contacts to which wires can be soldered. The third printing can be a resistor composition (DuPont 1711), fired at 850°C, or a conductor (DuPont 6120). If the resistor paste is used, the resistors can be trimmed after firing in order to balance the Wheatstone bridge. In most of our prototypes, conductor paste was used for simplicity of fabrication and to simplify analyses of initial sensor tests. The fourth pattern, forming the passivation over the reference legs, was usually printed and fired twice for added thickness. We had discovered that at concentrations of hydrogen above 4%, some hydrogen could enter the reference legs covered by only a single passivation layer. Several commercial and experimental passivation compositions were tested; a resistor encapsulant (DuPont 9137), fired at 525°C and a crossover dielectric (DuPont 9429), fired at 850°C, performed successfully. Because the components of the sensor are sintered at high temperatures, they are refractory and quite stable in air under ambient conditions, and give the device an inherent advantage for operation in harsh environments.

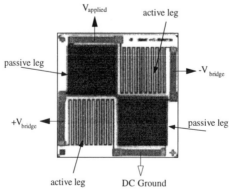

Figure 2. The ORNL hydrogen sensor.

Experimental Test Bed Descriptions

The Harsh Environment Test Bed (HETB). Some of the sensor tests were conducted at the HETB, an ORNL-fabricated stack/vent simulator consisting of two sections, a generator section and a stack section. The generator section is a horizontally mounted 2-inch (5 cm) stainless steel pipe through which gases and matrix modifiers, such as water vapor, may be introduced into the system. The generator section is flanged to the vertical stack section, a 4-inch (10 cm) diameter stainless steel pipe that contains twelve ports positioned at three different heights along the stack. Both the generator and the stack sections are equipped with sufficient heating to allow system operation to 250°C. For our tests, flows were controlled from two to 15 l/min depending upon the hydrogen concentration desired. The time constant or volume replacement time for this test configuration varied from 2 to 6 seconds, depending upon flow velocities. A commercially available blend of 4% hydrogen in argon was used to provide hydrogen to the system. Nitrogen was used as a diluent gas where necessary.

The test outline called for a series of sensor tests at three temperatures, (20°C, 90°C, and 200°C) matrixed against relative humidities of 0%, 80%, and 100%. However, not all of these moisture levels were obtainable at all stack temperatures. At 20°C and 90°C we were able to obtain arid, 40%, and 80% relative humidity and only arid conditions at 200°C. Power to the sensors was supplied by a Hewlett Packard, model 6205B dual dc power supply at 11 volts. Data acquisition was performed by a custom, PC-based system using National Instruments' LabView™.

Metrology R&D Laboratory (MRDL). A separate hydrogen sensor test bed at the MRDL facility was set up to perform the hydrogen concentration tests at concentrations to 30% in air and pressurized tests for simultaneous high humidity and temperature. In addition, it can perform the same tests as those at the HETB, which helps to identify testing flaws and instrumental error. Here, evaluation of the thick-film hydrogen sensor consisted of parametric tests to quantify sensitivity, time response, long and short term drift, reliability, and pressure/humidity sensitivity of the sensor under conditions representing some worst-case scenarios for accident conditions in a nuclear power reactor. Included within the matrix were high temperature conditions of 100°C, 150°C, and 200°C, combined with relative humidity requirements approaching 100% RH. To accommodate these conditions, testing pressures of approximately 260 psig (1800 kPa) were required to maintain near saturation conditions around the sensor. Precautions were required to ensure safe operation throughout these harsh conditions of the tests, while operating within the potentially explosive concentrations of 4 to 75%.

An easily reconfigured, explosion-proof gas flow loop was constructed to accommodate the wide range of conditions needed for these tests. A test chamber to house the sensor, and minimize system response time was constructed. This housing consisted of a short section of 5 cm stainless steel pipe, approximately 10 cm in length and capped on each end with a modified bell reducer. Overall housing volume was maintained at a minimum to maximize the thermal and pneumatic time response of the system. Plumbing connections to the test chamber were provided via 1/4 inch VCR (vacuum compression ring) components. Electrical connections, including a 9-volt dc excitation pair and a low level signal pair, were supplied through a high pressure/high temperature electrical feedthrough, comprising solid nickel-oxide clad copper wires embedded in a high temperature epoxy resin. Electrical connections to the sensor pad leads were initially established using a silver based, high-temperature solder. However, solder joint failures were often experienced at temperatures exceeding 175°C. These problems, combined with an overall difficulty in achieving a good solder joint to the sensor pad, led to the use of a high temperature, conductive epoxy for wiring connections of later tests.

Using the sensor housing chamber, a variety of test configurations were employed to evaluate the sensor under various conditions. Overall sensitivity and time response of the sensor was conducted using a flow-dilution configuration as presented in Figure 3.

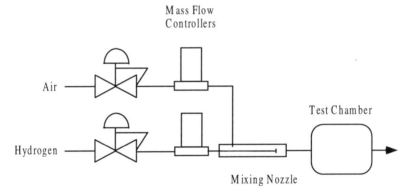

Figure 3. Sensitivity and step response configuration.

In this configuration, two mass flow controllers (MFCs) control the concentration of hydrogen in the flow stream. The mass flow controllers, STEC Model SEC-7330, are 1 l/min units calibrated respectively using air and hydrogen using a gravimetric method. To ensure adequate time response, the units were also checked for speed of response to a setpoint change against a high speed Honeywell Microswitch hot film device with a time response of approximately 10 ms. Time response of the MFCs was consistently better than 0.25 seconds. The time constant for delivery of test gas to this test chamber was calculated at 7 s. Overall accuracy was better than 0.5% of full scale flow. Gas mixing was optimized using an annular mixing nozzle. Flow rates were maintained at 1 l/min for all hydrogen concentrations to standardize the effect of flow velocity over the surface of the sensor. To evaluate the sensor at various temperatures, the test chamber was piped into a Precision Scientific Thelco oven with capability of temperatures from ambient to approximately 230°C, and stability of better than 2°C. Using this configuration, only dry, non-moisturized, flow streams were used.

Sensor tests were automated through National Instruments LabWindows™ and consisted largely of controlling and timing gas mixing sequences and data logging. Data acquisition and control were performed using a PC-based system with a National Instruments AT-MIO-16X analog input board, and an AT-AO-6/10 analog output board. Input data consisted of the excitation power voltage, the sensor output voltage, and several type-J thermocouple measurements. Analog outputs controlled the flow setpoints of each of the MFCs. The oven temperature was controlled via serial communication through the PC communication port.

Results and Discussion

Sensor Performance.

Stability. The output of the sensor in air with no hydrogen present varies little over time. HETB tests in the April and August periods included temperature ranges from 20 to 200°C, humidity conditions to saturation, and repeated exposures to hydrogen/argon/nitrogen/air mixtures between hydrogen concentrations of 0.5% to 4%. Figure 4 shows the outputs of two sensors, one that survived all of the temperature and humidity excursions, sensor H2(a); and the other, H2(b) that failed during high temperature tests on the April 9 date. Several other sensors placed in this test environment lost connectors when the solder failed at elevated temperatures, and stability measurements could not be made for them. More testing is required to fully characterize the sensor's stability.

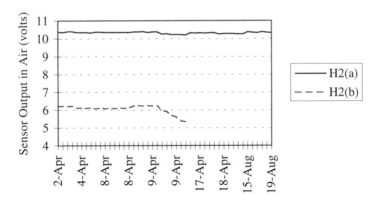

Figure 4. Output of two sensors at the HETB over a 4-month period.

Response time. Theoretically, the diffusion of hydrogen into a thin layer of palladium occurs on the order of milliseconds; in laboratory tests, however, the test set up can prevent direct observation of the actual response time of the sensor. Often the overall time response is the time constant of the test chamber, tubing, etc. The HETB time constant is between 2 and 6 s depending upon the flow rates dictated by gas mixing; at the MRDL facility, the time constant is 7 s. Other factors also have an effect. In Figure 5, we observe that temperature has little effect on the response time of a sensor in the HETB. For most of these points, the time to reach 90% of the maximum change in the hydrogen atmosphere is less than 30 seconds, regardless of concentration. All of the measurements used in this particular figure were made without the introduction of any moisture. There is a modest trend of slightly faster response at higher temperatures, as one would expect. Test data showed that the sensor output in the HETB begins to change around 4 to 10 s, which may be related to the time constant of the gas delivery system. In tests conducted 4 months apart, the time response increased for the later measurements, averaging about 100 s. The explanation may be that over time moisture or other constituents adsorbed on the active palladium influenced the normal kinetics of hydrogen gas sticking and dissociation. We have observed that after cycling the sensor through hydrogen a few times, the sensor did recover its original time response in the 20-30 s range. We are currently studying methods of eliminating the effect of adsorbed species by applying specialized coatings that are permeable to hydrogen, but not to moisture or other contaminants.

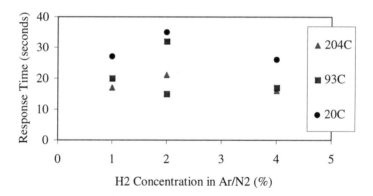

Figure 5. Time response of sensor H2(a) for three temperature regimes. The y-axis measures the time to reach 90% of the maximum sensor output for each exposure to hydrogen at the indicated concentration.

Response times for the same sensor at three humidity levels, 0%, 80%, and 100%, shown in Figure 6, indicate that the sensor response is somewhat slower as the relative humidity increases. (This test was conducted at the HETB. Test conducted at MRDL indicated no discernible difference in time response for different humidities. The test setup and procedure for each facility will be considered in analyzing these differences.) The time to reach 90% of the equilibrium response for each concentration is still less than one minute for all of the tests. Time responses for sensor recovery in air do not indicate any discernible trends. These initial tests are encouraging; nevertheless it will be important to understand more about the sensor response and life span under humid conditions, because humid conditions or conditions of varying humidity, are common to many industrial environments.

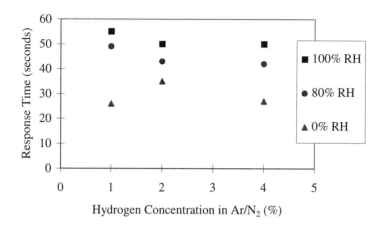

Figure 6. Time response of sensor H2(a) for three humidity regimes at room temperature. The y-axis measures the time to reach 90% of the maximum sensor output for each exposure to hydrogen at the indicated concentration.

All of the early tests were designed to introduce a rapid hydrogen concentration change in the test environment. However, because some processes exhibit a gradual build up of gas, such as the evolution of hydrogen from overcharged lead acid batteries, future tests are planned to determine the sensor's stability and accuracy as hydrogen gas concentration changes slowly.

Linearity. Early functional tests of the sensor promised good correlation between sensor output and hydrogen concentration. Figure 7a shows the responses for a sensor repeatedly exposed to a range of concentrations of hydrogen in air (0.5 to 3%), at room temperature and humidity. The linear fit across the corresponding range of concentrations is shown in Figure 7b. Another sensor was tested through the range of 0.5% to 30% H_2 in N_2, Figure 8. Although we suspect the sensor behavior will be different for air as opposed to an inert environment such as nitrogen, we do not have sufficient data at this time to support any conclusions. A promising result, however, is that even at 30% hydrogen, the sensor does not appear to saturate. Noticeable in Figures 7a, 7b, and 9 are what appears to be low sensitivity of the sensor to concentrations at or below 0.5% H_2. Previous test results using different test apparatus have shown that the sensor is still quite sensitive at this level of concentration. We have since verified that this was a limitation of the MFC supplying the hydrogen, which was selected for time response rather than accuracy at the lower concentrations. Future tests will use bottled premixed hydrogen in selected matrix gas such as air, nitrogen, or argon so that we can more accurately meter lower concentrations of hydrogen. Nevertheless, the linearity of the sensor over gradually stepped concentrations between 0.5 and 17% is surprisingly good, as shown in Figure 9.

98

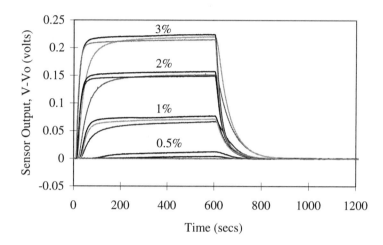

Figure 7a. Step response to several H_2 concentrations in air.

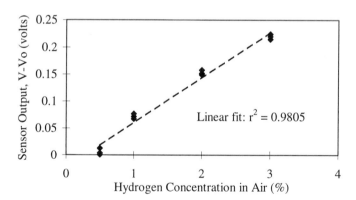

Figure 7b. Linear fit for curves in Figure 7a. Each point represents the maximum sensor output for exposure to indicated concentrations.

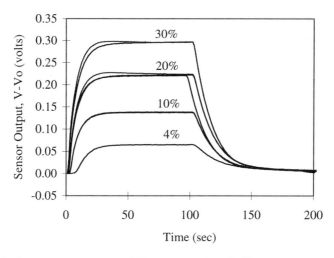

Figure 8. Step response to several H_2 concentrations in N_2.

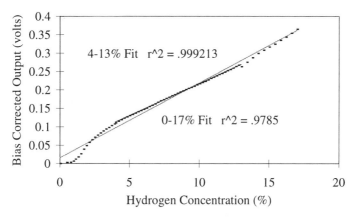

Figure 9. Linear fits for sensor response to gradual increases in concentrations between 0 and 17% H_2 in air.

Elevated Temperatures. As temperature increases, the amount of hydrogen dissolved in Pd at a given hydrogen partial pressure decreases. This is demonstrated for several sensors in Figure 10. Obviously, if the sensor is to be used quantitatively, appropriate calibration curves will have to be developed for the intended sensor environment. In addition, packaging design may incorporate on-board temperature measurement to compensate for temperature influence on sensor output.

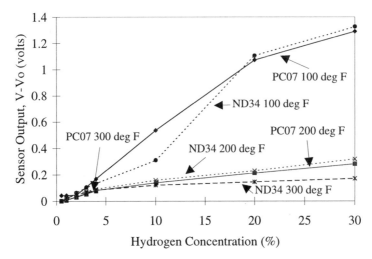

Figure 10. Relationship of sensor output to H_2 concentration in air at three temperatures.

Packaging Issues. Packaging design will be inevitably driven by the intended application domain. For high temperature use, connections to the sensor must be secure and reliable. Most solders will not perform well above 200°C. We are experimenting with conductive epoxies, but are aware that wire-bonding attachments are probably the most promising long term approach.

We suspect that over time, moisture adsorbing to the sensor surface will be detrimental to the sensor response time in humid environments. We have proposed covering the active palladium legs with Teflon or other hydrogen-permeable materials to exclude molecules of water and other possible contaminants. The covering may also slow the response rate, so functional tests of such modifications must be performed for each intended application.

In a current project, we are designing a sensor layout for a 160-mm^2 (0.25-in.^{2}) substrate. This size is convenient for commercial packages and connector configurations used by some sensor manufacturers. Ideally, the sensor can be mounted in a commercial package or monitor housing, and accompanying off-the-shelf electronics circuits can be tuned or programmed to operate the new sensor.

Conclusions

We have demonstrated that the thick film sensor design is sturdy and responsive to hydrogen under the temperature and humidity tests described in this report. Our failures with solder connections at temperatures of 175°C have shown us that any sensor package design for high temperature use will have to also incorporate suitable materials and connections for reliable operation. The sensor has a linear response to hydrogen in matrix gases of air, nitrogen/air, and argon/nitrogen/air for concentrations between 0.5 and 30%. We would like to conduct additional tests to verify repeatability and accuracy. Also, sensor performance differences, such as time response and sensor output in air versus inert environments would be interesting to study. Future tests will address the long term stability of the sensor output and sensor materials; the upper and lower limits of response; and operation in additional humidity and temperature regimes. For reliable operation in industrial environments, it will be important to characterize sensor drift and life time.

By using thick film techniques, many design changes can be easily made to accommodate requirements for specific applications. However, it will be important to understand the limitations of the thick film components and fabrication as they relate to sensor miniaturization, power consumption, and sensor sensitivity to hydrogen gas. Also if materials, such as Teflon, are to be used to protect the palladium metallization from adsorbed water or other contaminants, fabrication practices will have to be devised and then evaluated for economic practicality. Additional fabrication steps, especially for incorporation of non-standard thick film materials, will likely increase the cost of the sensor. Alternatively, there may be thick film materials that could be developed to provide the same function.

Literature Cited

1. Lundström, I. "Hydrogen-Sensitive MOS Structures. Part 1: Principles and Applications," *Sens. Actuators* **2**, 403-26,1981/82.
2. Lundström, I "Hydrogen-Sensitive MOS Structures. Part 2: Characterization," *Sens. Actuators,* **2**, 105-38, 1981/82.
3. Lechuga, L. M.; Calle, A.; Golmayo, D.; and Briones, F. "Hydrogen Sensor Based on a Pt/GaAs Schottky Diode," *Sens. Actuators B*, **4**, 515-18, 1991.
4. Shaver, J. "Bimetal Strip Hydrogen Gas Detectors," *Rev. Sci. Instr.,* **40**, 901-5, 1969.
5. Godwin, J.; and Klint, R. V. "Hydrogen Detector System," U.S. Patent 3,927,555, 1975.
6. Butler, A. "Optical Fiber Hydrogen Sensor," *Appl. Phys. Lett.,* **45**, 1007-9, 1984.
7. Hughes, R. C.; Buss, R. J., Reynolds; S. W., Jenkins, M. W.; and Rodriquez, J. L. *Wide Range H2 Sensor Using Catalytic Alloys*, SAND-92-2382C, Sandia National Laboratory, Albuquerque, NM 1992.
8. Lewis A. *The Palladium Hydrogen System*, Academic Press, NY, 1967.
9. Hoffheins, B. S.; Lauf, R. J.; Nave, S. E. "Solid-state, Resistive Hydrogen Sensors for Safety Monitoring," *Instrumentation, Controls, and Automation in the Power Industry,* **36**, Proc. Third Int'l Joint ISA and POWID/EPRI Controls & Instrumentation Symposium, Instrument Society of America, Research Triangle Park, NC, (1993) pp.189-203.
10. Hoffheins, B.S.; and R.J. Lauf "Thick Film Hydrogen Sensor," U.S. Patent No. 5,451,920, 1995.
11. Felten, J. J. "Palladium Thick Film Conductor," U.S. Patent No. 5,338,708, 1994.

Chapter 9

Gas Sensors for Automotive Applications

L. Rimai, J. H. Visser, E. M. Logothetis, and A. Samman

Ford Research Laboratory, MD 3028, P.O. Box 2053, Dearborn MI 48121–2053

Sensing certain components present in the exhaust gas stream of engines has become a very important application for chemical sensor technology. The use of catalytic electrode ZrO_2 Nernst cells as oxygen sensors for stoichiometric control of engine operation is standard in production automobiles. In this paper we will discuss different sensor structures that may fulfill other important exhaust gas sensing requirements. Some of these involve the use of ZrO_2 cells in the oxygen pumping mode of operation, and are useful in extending oxygen sensing capabilities over a wide concentration range, as well as making it possible to sense other constituents such as hydrocarbons, with some degree of specificity. Gas composition sensing mechanisms using Si based microcalorimetric technique and SiC based metal/oxide/semiconductor capacitors will also be discussed, as these have good potential as hydrocarbon sensors.

Since the late sixties the reduction of emissions has been of importance as a driving force in the development of automotive technology. The use of catalytic exhaust processing combined with controlled stoichiometric engine operation allowed the production of vehicles in compliance with the emission standards established by legislation at that time in the United States. By 1990 exhaust emissions were at levels of 0.4 grams/mile hydrocarbons (4% of previous unregulated levels), 7 grams/mile of CO and 0.4 grams/mile of nitrogen oxides (25% of unregulated levels)(1). In the United States, new legislation enacted in 1990 considerably tightened the regulations, with a progressive time table to be implemented over a 10 year period. California has established even more stringent emission limits, requiring for example that in the 1997 model year a substantial fraction of the vehicles emit less than .075 g/mile hydrocarbons, 3.4 g/mile CO and 0.2 g/mile nitrogen oxides, with another factor of 2 reduction for a smaller fraction of the vehicles. In addition, the mileage requirement for maintaining the limits on emissions was extended, from 50000 to 100000 miles (or

©1998 American Chemical Society

10 years). There are also limits on "evaporative emissions", i.e. emissions from the fuel system. It is expected that a number of other states and eventually even the federal government might follow the California example. As hydrocarbon and CO exhaust emissions result from incomplete combustion of the fuel charge admitted into the engine cylinder, improvements in combustion efficiency might lead to emission reductions. However, as the untreated emission levels from a well operating engine already correspond only to a small fraction of the initial fuel charge, and considering the complicated, highly nonlinear and unstable nature of the in cylinder combustion process, such improvements are difficult and their implementation can easily lead to diminishing returns. Thus the development of catalytic emission treatment and external engine control were spawned.

As limits are set for both, reduced (Hydrocarbons, CO) and oxidized (NO_x) species, the catalyst has to operate in a manner to bring to equilibrium the oxidation of the former to CO_2 and H_2O, as well as the reduction of the latter to N_2 and O_2. This requires operation in appropriate ranges of temperature and composition. The temperature range is realized in the exhaust stream, under most engine operating conditions (400 to 800 °C) where the kinetics in the presence of a suitable three way catalyst (as it is called) are sufficiently fast. For both, the reductive and oxidative reactions to be efficient, the charge composition has to be within a small range of exact stoichiometric equivalence ratio $\lambda = 1$. This equivalence ratio λ is defined as the actual oxygen to fuel ratio normalized to its equilibrium value corresponding to exact stoichiometry.(1). If the temperature condition is not satisfied, as in the case of cold start, the reaction kinetics will be too slow, and emissions will be high. For most other driving situations the catalyst will be operational, and emissions can be kept below the limits set by the pre 1990 legislation, with the use of the "stoichiometric" engine control technique. This technique requires the use of an exhaust gas sensor to transduce the equivalence ratio for the charge into an electrical signal that can then be used to adjust appropriately the amount of fuel injected into the cylinders to preserve near stoichiometric operation. This is achieved by measuring the oxygen concentration in the fully oxidized exhaust stream ahead of the three way catalyst, as this oxygen concentration has a very rapid dependence on the equivalence ratio exactly in this region near $\lambda = 1$ (1).

The zirconia Nernst cell

The sensor most often used to determine this equilibrium oxygen concentration in the exhaust gas is based on a Y_2O_3 stabilized cubic ZrO_2 electrochemical cell with porous platinum electrodes. The presence of the stabilizing impurity entails the presence of oxygen vacancies within the material, which become positively charged as the corresponding electrons remain bound to the Y donors (2). At temperatures above a few hundred degrees the mobility of these positive oxygen vacancies becomes sufficiently large that the material behaves as a good ionic conductor. The electrons exhibit negligible mobility, as most remain bound even at these temperatures. Under open circuit condition the total current density through a layer of the ZrO_2 is zero:

$$j = q_v(D_v\nabla c_v + \mu_v c_v \nabla \phi) + q_e(D_e\nabla c_e + \mu_e c_e \nabla \phi) = 0 \qquad (1)$$

The diffusion constants D_α and mobilities μ_α are related by the Einstein relation

$$D_\alpha = kT\mu_\alpha / q_\alpha, \tag{2}$$

c_α, q_α, are the concentration, charge of species α (oxygen vacancies v and electrons e), k is the Boltzman constant, T the absolute temperature, ϕ the electrostatic potential and $q_v=-2q_e>0$ (2). In a one dimensional approximation and neglecting the mobility of the electrons, equation 1 integrates to yield the Nernst relation between the oxygen vacancy concentrations at opposite ends of a ZrO_2 slab and the corresponding open circuit potential difference:

$$. \quad \phi_2 - \phi_1 = (kT / q_v)\ln(\frac{c_1}{c_2}) \tag{3}$$

Under steady state conditions the concentration of oxygen vacancies in the ZrO_2 near the surface should be proportional to the concentration of these vacancies on the surface. Furthermore, the oxygen from the gas phase which is adsorbed onto the platinum as atomic oxygen will diffuse rapidly through the porous Pt to the ZrO_2 surface. The oxygen vacancies in the zirconia recombine with this adsorbed atomic oxygen. In a reducing gaseous atmosphere the opposite will happen. Oxygen vacancies will be formed by the reduced species reacting with the oxygen within the oxide. Thus in steady state, a variation of the adsorbed oxygen surface concentration will entail a variation in the opposite direction of the vacancy concentration in the ZrO_2 near the surface. If the Pt film were a fully effective catalyst then it would keep the gaseous mixture near its surface, in thermal equilibrium. This is the so called transport limited situation when the rate of reaction is determined by reactant diffusion. This oxygen concentration, corresponding to equilibrium, will exhibit the characteristic sharp step somewhere in the vicinity of $\lambda = 1$, which will, in turn, be reflected in the near surface vacancy concentration (2,3). When one electrode of a Nernst cell is in contact with a constant reference oxygen concentration whereas the other is in contact with the exhaust stream, it can be seen from equation 3 that the cell's open circuit voltage will also exhibit a step as the equivalence ratio crosses the value $\lambda = 1$. The difference between the measured cell voltage, and its value corresponding to the middle of the transition region provides an error signal to be used in the engine control feedback loop. When the assumption of equilibrium in the gas near the electrode does not hold, for example if the temperature of the cell is too low, a step in the output voltage is still observed, but the corresponding value of λ will be dependent on reaction kinetics (2).

Oxygen pumping sensors

Even under local equilibrium conditions the open circuit Nernst cell is only useful to determine the stoichiometric point, as its output is essentially constant below and above it. Measurement of the equilibrium O_2 concentration in the exhaust, away from $\lambda=1$ is of importance if lean engine operation is considered. Under lean conditions hydrocarbons and CO in the untreated emissions tend to be lower provided combustion stability is maintained. To maintain the latter, the exhaust O_2 concentration signal could be used as a control variable. Such a signal can be obtained using the same ZrO_2 cell in the "oxygen pumping" mode (3,4). This is an important mode of operation of zirconia sensors which also might lead to practical combustible gas sensor structures. As seen in equation 1, if a current is driven through the ZrO_2 cell, the voltage measured between the electrodes will be shifted from the value of equation 3. This voltage shift corresponds to an electric field inside the ZrO_2, $E=j/\sigma$. For a one dimensional system, in steady state j is independent of the spatial coordinate x. To a good approximation the carrier (O_2 vacancy) concentration is constant through the bulk of the zirconia (2). Thus the conductivity σ as well as the electric field should also be constant. The voltage shift corresponds to a potential drop across the internal resistance $R=d/\sigma s$ of the zirconia slab, where d and s are its thickness and cross section. The voltage between the electrodes will be

$$V = RI + ({kT}/{q_v}) \ln(\frac{p_1}{p_0}) \tag{4}$$

where the resistance R also includes a term to account for the finite resistance of the porous electrodes. I is the current and p_1, p_0 are the oxygen partial pressures near the opposite electrodes. The charge carriers being the O_2 vacancies, the passage of current implies transport of these vacancies from the vicinity of electrode 0 to the vicinity of electrode 1, with consequent inverse transfer of $O_2^=$ ions. If the device is immersed in an atmosphere where the O_2 partial pressure is p_0, and electrode 1 is covered by some sort of a diffusion barrier which will maintain a concentration gradient proportional to the traversing oxygen flux, then the steady state current is related to this flux by

$$I = 2q_v AD(\frac{p_0 - p_1}{kT}) \tag{5}$$

where D and l/A are respectively the oxygen diffusion constant and effective diffusion length appropriate for the diffusion barrier. Equations 4 and 5 yield the I-V characteristic of the device, which is represented by the implicit relation (4)

$$V - IR = ({kT}/{2q_v}) \ln(1 - \frac{IkT}{2q_v ADp_0}) \tag{6}$$

At low currents the device behaves as a resistance R. Equation 6 implies that there will be a limiting current that can be forced through the device, $I_S=2q_vADp_0/kT$, corresponding to total depletion of oxygen at electrode 1. Attempting to force progressively larger currents through the device by imposing larger voltages will eventually cause rapid current increases associated with electrolytic decomposition of the ZrO_2. Also, if the gas does contain oxidized species, such as water or carbon dioxide, current steps can occur at set voltages due to their electrolytic decomposition.

Suitably designed cells can actually be used in this manner to sense the presence of such oxidized species (3,4,5).

One of the most important applications of this pumping technique is in a dual cell structure shown schematically in Figure 1, which is the basis for the "universal" oxygen sensor (3). The small common volume Ω in between the two cells is connected to the outside through a diffusion barrier which ensures that a concentration gradient proportional to the corresponding flow rate will be maintained across it for the various components of the gas. At sufficiently high oxygen partial pressures p_0 outside Ω compared to that of the other reduced species (lean conditions) and assuming that thermal equilibrium prevails inside Ω, the flow rates will be given by

$$F_0 = AD_0(p_0 - p_\Omega) / kT \qquad (7)$$

for oxygen and

$$F_j = AD_j p_j / (kT) \qquad (8)$$

for species j. D_j, p_j are respectively, the barrier diffusion constant and partial pressure and p_0, p_Ω are the partial pressures of oxygen outside and within Ω. If a pumping current I is forced through one of the cells, and the open circuit voltage V is measured for the other, the device will have the following I-V transfer function

$$I = (\frac{2q_v A}{kT})\{D_0 p_0[1 - \exp(-\frac{2q_v V}{kT})] - \sum_j n_j p_j[D_j - D_0 \exp(-\frac{2q_v V}{kT})]\} \qquad (9)$$

n_j is the number of moles of species j oxidized by one mole of O_2. Equation 3 was used for the open circuit of the second cell, replacing the O_2 vacancy ratio by the inverse oxygen partial pressure ratio as discussed in the preceding section (6). Equation 9 shows that by using the device in a constant voltage feedback mode one can obtain, for arbitrary currents, a linear response of pumping current as a function of O_2 concentration for example(6,7). This feature makes the dual cell oxygen pumping device a useful exhaust gas oxygen sensor under lean conditions (whereas the Nernst cell having only a logarithmic response is insensitive except very close to stoichiometry). This has been the major application of this device, although it can also be used as a combustible sensor. If for instance, in equation 9 V is set equal to 0, the response becomes independent of the O_2 partial pressure and depends only on a suitable linear combination of concentrations (partial pressures) of the various reduced species present. The weighting in this linear combination depends on the relative values of the various diffusion constants. A series of measurements of current for a set of suitably chosen voltages can be analyzed to yield detailed information on the chemical makeup of the gas mixture. Equation 9 also shows that the contribution of any specific component can be eliminated by a specific choice of V. This might be especially useful for exclusion of methane, (or CH_4 and CO by using paired measurements) as is illustrated in Figure 2 with experimental I-V curves obtained for such a device using methane/air mixtures. The experimental sensitivity of this same device to different gases in air is shown in Figure 3, for two preset voltages of the Nernst cell (7). As negative voltage is increased in magnitude relative to kT/q_v, the response $I(-\infty)$ tends to become proportional to the equilibrium oxygen concentration

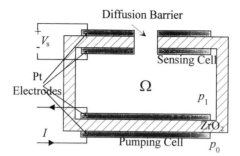

Figure 1. Schematic representation of a dual cell oxygen pumping electrochemical device.

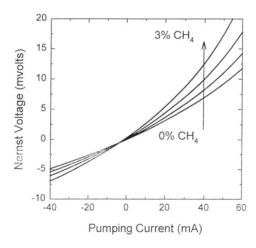

Figure 2. Current Vs Nernst voltage characteristics of a dual cell zirconia oxygen pumping device for 0, 1,2 and 3% CH_4 in 100 sccm of flowing air. The cell temperature is 1000 °K. The curves cross at a single point, corresponding to the condition when the sensor is insensitive to methane.

$(p_0 - \sum_j n_j p_j) / kT$ as the exponentials dominate the right hand side of equation 9; this is the equilibrium oxygen concentration corresponding to the equivalence ratio of the initial charge in the cylinders.

In what we have seen so far, the zirconia sensor technology can be used to determine when the equivalence ratio of the engine charge corresponds to stoichiometry. During lean operation, it can provide information about the actual value of the equivalence ratio. Also, for gas mixtures with excess oxygen it will provide information about the composition in terms of reduced species (8). There are sensor technologies other than those based on ZrO_2 that will yield information about such combustibles present in lean gas mixtures, and with controlled addition of oxygen to the vicinity of the sensors they can also provide such information for rich gas streams. We will discuss two such sensors in the remaining section of this paper: the microcalorimeter will be described rather briefly and the SiC based catalytic gate metal/oxide/semiconductor (MOS) capacitor (9) which will be treated in a little more detail. We should note that all these sensors involve catalyzed chemical oxidation of the sensed species and therefore do not have inherently selectivity. Only spectroscopic processes can be fundamentally species selective. However, with the possible exception of non dispersive IR CO sensor, their on board automotive applications are considerably further beyond the horizon than the devices covered in this paper.

The Silicon Based Microcalorimeter

Of all possible techniques of gas composition sensing in reducing-oxidizing mixtures, calorimetry is the simplest in terms of the transduction mechanism. The sensor consists of a catalyst film deposited onto a low mass support which also holds a temperature sensor. The device is held in the gas stream by a heat insulating structure. The temperature of the sensor should be sufficiently high to activate the catalyst and enable the oxidation of the combustibles in its vicinity. The oxidation reactions being exothermic, the rate of excess heat generated near the surface of the catalyst should be proportional to the combustible concentration at mid-stream, provided the oxidation rate is sufficiently large that the combustible concentration near the surface of the catalyst is negligible (mass transport limited condition). The requirement that there be excess oxygen concentration is implicit. In steady state, the reaction will cause a rise in the temperature of the sensor which is determined by the balance between the rate of heat generation due to the chemical reaction and the corresponding increase in the rate of heat loss. If the latter is assumed to be a linear function of the temperature, as is rigorously the case for conduction and convection losses, the temperature rise will be proportional to the midstream combustible concentration c:

$$\Delta T = Q_0 Dc / hb \tag{10}$$

where Q_0 is the heat of reaction (per molecule), D is the combustible diffusion constant, b is an effective diffusion boundary layer thickness, and h is an effective heat transfer coefficient representing all losses (10,11). As can be see from equation 10, for mixtures of different combustible species the response of this device will be proportional to a linear combination of the individual concentrations with the weighting factors being the diffusion constants (see also equation 9). The base temperature of the device has to be in the range of 300 to 500 °C and is usually attained by resistive heating. As the reaction rates are temperature sensitive, the device has to be operated either in a constant temperature feedback mode or in a differential system with two devices close to each other in position, differing only in that the reference device has no catalyst (10). Calorimetric devices have been used commonly for the detection of combustibles at concentrations near explosive limits (11), but without the response, sensitivity and precision mass manufacturability required by automotive applications. Si micro machining and IC technology would enable the fabrication of such devices with sufficiently small thermal mass to yield sub second response times and with sensing elements adequately insulated to increase sensitivity. This manufacturing technology is also inherently geared for the large scale efficient production which would enable their practical use in automotive applications. Figure 4 shows experimental data obtained from such a differential micromachined calorimeter in a laboratory gas flow system (12).The calorimeter consisted of a pair of thin film platinum resistance thermometers deposited respectively on a pair of low conductivity micromachined membranes and covered by a thin passivation film. A layer of catalyst was deposited on top of only one of the membranes, the temperature rise being derived from the differential resistance measurements. The results so far seem to indicate that for a number of hydrocarbons, CO, and H_2, the transport limit assumption indeed applies, as the response seems to be independent of the specific catalyst used and the response ratios correspond to the ratios of the reaction heats. However, in some cases (saturated hydrocarbons), this approximation fails, as the response becomes catalyst dependent (12). For finite rate first order oxidation, the proportionality coefficient between the heat evolution rate and the midstream concentration is reduced by a factor $[1 + (D/br)]^{-1}$ where r is the surface reaction rate coefficient. This causes the device sensitivity to be dependent on the catalyst chemistry.

As we saw in all discussions above, the ideal operation of all sensor types always requires attainment of equilibrium in an oxidation reaction ahead of the generation of the electrical output signal. Under conditions where the response depends on reaction rates, sensing will still take place, but reliable response will require rather complicated referencing procedures. Equilibration <u>sometimes</u> (12) can be enhanced by driving the reaction with excess oxygen, which can be done in an easily controllable way by the oxygen pumping procedure discussed in section 3, or by increasing the operating temperature. This latter procedure combined with the recent availability of reasonable quality wide band gap semiconductor, namely SiC, opened up the possibility of the use of catalytic gate MOS type structures for sensing of automotive exhaust gas components of relevance to emission control.

The SiC based MOS sensor.

Catalytic gate MOS structures based on Si can function as hydrogen sensors (13,14). They were also shown to respond to certain hydrocarbons but the kinetics were by far too slow for useful applications due to limit on operating temperature imposed by the use of Si (15). A Si MOS sensor was first developed by Lunström and coworkers (16)

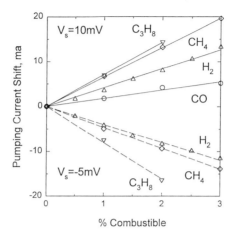

Figure 3. Response of the dual cell pumping device to various combustibles in air for 2 set voltages V_S of the Nernst cell. The linear dependence of the current on combustible concentration is clear even for these low currents. The ordinate is the shift in pumping current required to keep V_S constant.

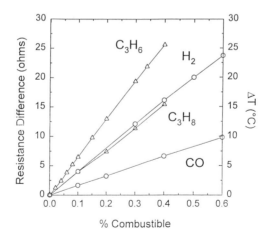

Figure 4. Response of a differential microcalorimeter to combustible concentration; difference in resistance (left scale), and the corresponding temperature difference (right scale). The gas flow velocity was 1 cm/s, the temperature in the non catalytic element 360 C and the resistance vs. temperature slope was 3000 ppm/ °C

and the literature is reviewed in the publications by Lundström and Baranzahi (17,18). Since 1989, there has been research on similar sensors based on SiC, mostly at Linköping University (19,20,21). SiC based sensors will operate at higher temperatures and respond not only to H_2, but also to a number of hydrocarbons. It has also been shown that they do respond to composition changes in the exhaust of an automobile engine (22). In the following we present a simplified discussion of the possible mechanisms involved in the functioning of these sensors, and include some experimental results that we have obtained using Linköping devices. We are limiting our considerations to the capacitors as the sensing process takes place in these structures, even when they are incorporated as gates to field effect transistors (19). The use of FETs based on SiC has been difficult due to low mobilities, especially in the near surface channels. The generation of the output signal in the capacitive devices does not depend directly on the carrier mobility in the semiconductor. Actually when the gas response signals are obtained from an MOS capacitor operated in the depletion mode or from a reverse biased Schottky diode, the carrier mobility in the semiconductor only affects the electrical response time, which is estimated to be well below that for the catalyzed chemistry or for the gas transport.

The simplest model for an MOS structure, shown schematically in Figure 5, consists of two capacitors connected in series. The capacitance C_i is formed by the insulating layer (usually SiO_2) between the gate metal and the semiconductor. In the ideal case this insulator insures that no DC current flows through the device. The other capacitance represents the layer of semiconductor near its interface with the insulator where, upon application of a gate voltage, charge redistribution takes place (depletion, inversion). For a perfectly insulating oxide, even with a DC voltage applied, the system will be in equilibrium, as no current can flow through it, which means that the Fermi level must be constant throughout the system. Because work functions of the metal and the semiconductor are different there will be an equilibrium voltage shift between the semiconductor interface and its bulk. As the carriers in the semiconductor are mobile this entails a non zero charge density near the surface, namely carrier accumulation or depletion, depending on the sign of this shift. If an additional voltage V is applied between the gate and the semiconductor bulk, the charge distribution within this region, which now supports an electric field, will be further modified. The potential at the oxide/semiconductor interface as well as the width of this charged region change with the applied voltage. At every point x in this "active" region, the net charge density ρ which is proportional to the difference between electron and donor concentration (for n type) is a function of the potential $\varphi(x)$ via the electron distribution function. This potential on the other hand is related to said charge density by Poisson's equation

$$\nabla^2\varphi(x) = -\rho(\varphi(x)) / \varepsilon = d^2\varphi / dx^2 \qquad (11)$$

where ε is the dielectric constant of the semiconductor and x is the spatial coordinate along the normal to the interface in a one dimensional approximation. Solution of the Poisson equation yields a relationship between the potential at this interface, φ_0, and the electric field on the semiconductor side of this same interface, $E_0 = -d\varphi/dx|_{x=0}$. As this field is proportional to the interface charge density, $Q_0(\varphi_0)=\varepsilon E_0(\varphi_0)$, the small signal capacitance of the active layer will be given by the following derivative

$$C(\varphi_0) = \frac{dQ_0(\varphi_0)}{d\varphi_0} \tag{12}$$

The solution of equation 11 for the insulator yields the dependence of φ_0 on V:

$$\varphi_0 - V = -\frac{1}{\varepsilon_i} \int_{-\delta}^{0} dx \int_{-\delta}^{0} dy \rho_i(y) + \frac{\delta Q_S}{\varepsilon_i} \tag{13}$$

where $\rho_i(x)$ is the (fixed) charge distribution through the insulator and ε_i its dielectric constant. Q_S is a surface charge on the insulator capacitance $C_i=\varepsilon_i/\delta$. The integration is over the thickness δ of the insulator, with the semiconductor interface taken at $x = 0$. The small signal total device capacitance C_T is given by

$$C_T(V + \Delta) = \frac{C(V + \Delta)C_i}{C(V + \Delta) + C_i} \tag{14}$$

where Δ represents the right hand side of equation 13. In the one dimensional approximation the capacitances are defined per unit area. A change in the charge distribution either throughout the insulator, affecting the integral in equation 13, or on its surfaces, affecting Q_S, will cause a change in Δ. Thus, a change in the steady state charge distribution in the insulator will simply entail a shift of the device's $C_T(V)$ curve along the voltage axis. This is the essence of the sensing mechanism, namely that the presence of a certain molecular species in the atmosphere near the gate causes a change in the distribution of charges on the insulator side of the device, which in turn is reflected in the shift of the $C_T(V)$ curve. Naturally if the gas composition of the atmosphere changes, in order to establish the new charge distribution, a transient current must flow, and the corresponding time dependent response cannot be discussed in terms of such a simple model. Also, the situation becomes considerably more complicated if losses in the insulator have to be accounted for, except when the device is operated under depletion conditions, i.e. when there is a finite region in the semiconductor where no charge carriers are present.

Figure 6 shows typical traces of capacitance and conductance versus gate voltage obtained with a Linköping (19) device. The data were recorded at a frequency of 1 MHz, with an HP4192A Impedance Analyzer. The device was held in a ceramic structure positioned horizontally on top of a resistive heater, inside a small flow tube. The back contact was made through a Pt foil while a Pt wire with a small sphere at the tip made contact to the gate. This front contact was held down by a small weight. The device was held, in this particular case, at 512 °C and the total gas flow rate was 1000 sccm. The combustible (propane in this case) concentration dependent shift is clearly seen from the traces. The capacitance and conductance were calculated from an equivalent parallel RC circuit model. This data shows the horizontal shift of the capacitance curve, as discussed in the model above, but it also indicates, by the presence of finite conductance, that this simple picture does not provide a complete

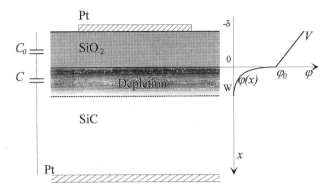

Figure 5. Schematic model of a MOS capacitor. The curve on the right illustrates the variation of the electrostatic potential in the ideal case of zero current through the device.

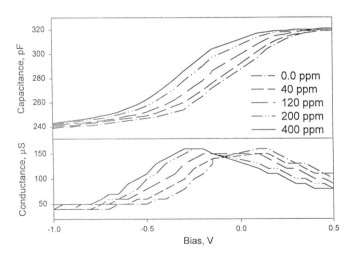

Figure 6. 1 MHz. small signal admittance measurements on a Linköping SiC based MOS capacitor, as a function of gate bias. This admittance is modeled as a parallel RC circuit, capacitance C on top and a conductance G=1/R on the bottom. The device is at 512 °C in a 1000 sccm flow of N_2 with 200 ppm O_2 and C_3H_8 as indicated for the various curves.

description of the processes ongoing in the device. The peak in the conductance G near zero bias for the curve taken in the absence of the combustible shifts rigidly with the capacitance curve. In the absence of combustible its height decreases with decreasing frequency with a corresponding shift in the bias in a manner consistent with the presence of carrier traps in the semiconductor in the vicinity of the interface (9). However, the rapid increase of the background loss as we move from depletion to accumulation seems likely to be related to losses in the oxide. Normal high frequency C(V) curves can be obtained for these devices, in the temperature range of 250 to 550°C, down to frequencies as low as 100 kHz as long as the bias remains within a relatively controlled range, specifically by remaining slightly shy of full accumulation. Below 100 kHz the bias range where stable characteristics can be obtained becomes rather small. This again might be the reflection of losses in the oxide, which are probably due to motion of massive carriers such as oxygen vacancies for example.

A specific mechanism has been used to explain the hydrogen sensing process in these devices (16,17). Hydrogen sensitivity seems to be strongest for Pd gates, where hydrogen dissociates on the surface and in its atomic form easily penetrates through the relatively thin Pt film (100 nm). The extra charge in the insulator that would account for the voltage shift of the C(V) curve might be the result of physical separation between the proton and the electron. Same mechanism would also explain the hydrogen sensitivity when Pt gates are used, as hydrogen dissociation probably also takes place on Pt, and there might be sufficient porosity in the thin gate films for transport of this species to the insulator. The hydrocarbon sensitivity of these devices can then be ascribed to hydrogen generated as an intermediary in its Pt catalyzed oxidation (18). There have been some arguments in the past, based on rather limited experimental evidence that have cast some doubt on this very simple hydrogen driven mechanism (15). Still, the discussion above shows that any neutral intermediate or product of the catalyzed oxidation, that can penetrate thorough a suitably porous gate and ionize at the interface or within the oxide can induce the same type of sensor response. In insulators the electron mobility is negligible due to very narrow bands, very wide band gaps and the possibility of deep laying electron traps. However, in many cases, at elevated temperatures cations may have a relatively large mobility, as for example in cubic defective ZrO_2, entailing charge separation. Especially in oxide insulators this might involve oxygen vacancies and atomic oxygen in the reacting gas. This type of mechanism would also explain combustible sensor action in such MOS devices.

Figure 7 shows the response of another Linköping device (20) to propane at constant bias by recording the capacitance changes due to stepwise changes of the combustible concentration in the gas at an oxygen concentration of 200 ppm in a 1000 sccm flow of nitrogen. Figure 8 shows data obtained for both propane and propylene at various oxygen concentrations, the abscissa being the equivalence ratio (stoichiometry corresponds to 1). These are steady state data and were obtained as that in Figure 7 at the end of each 5 minute lasting step in combustible concentration. Due to the occurrence of small drifts, the data was taken for each concentration cycle with respect to its initial zero concentration value. The results indicate that below and above stoichiometry the sensitivity to concentration decreases for all values of oxygen concentration, at least hinting that in analogy with the ZrO_2 devices these are

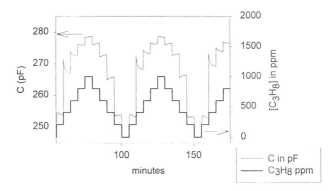

Figure 7. The variation of the capacitance (thin line) of a SiC based MOS device as a response to a stepwise variation of the propane concentration (thick line) in a 1000 sccm flow of N_2 with 400 ppm O_2, 0 V gate bias.

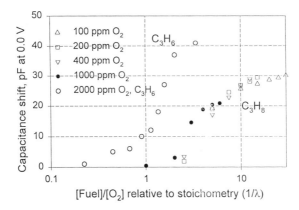

Figure 8. Capacitance shift at zero bias from its value in the absence of combustible, as a function of the inverse of the equivalence ratio, for C_3H_6 in 2000 ppm of O_2, and for C_3H_8 at various O_2 ppm in 1000 sccm N_2. The device was held at 512 °C

equivalence ratio sensors. As the data of Figure 7 shows, the transient response after each concentration step has a fast component and a much slower component. Estimates based on the characteristics of the flow system indicate that the fast component is determined by fluid mechanical convection and transport delays. The slow component might be related to low mobility charge transport in the insulator, thus indicating where one of the important developments in the technology of these devices has to occur in order to make them practical for automotive applications.

Concluding Remarks

The previous sections presented a review of some technologies that are at present either being applied, or are under some level of consideration for applications in control and reduction of automotive emissions. Their common feature is the chemical basis of the sensing mechanism, namely the quantification of some important types of active constituents of the exhaust stream by assessing its reduction/oxidation state. As such, they are either oxygen or combustible sensors. As combustible sensors, they do not have inherent selectivity to the different reduced species in the exhaust but, as implied in the discussions above, their response is expected to be different for species that have markedly different oxidation rates. Presently existing emission regulations already have some selectivity requirements, as CO and hydrocarbon emission limits are separately specified, with the latter excluding methane. The chemistry in all these devices is catalytic, and some level of specificity might be obtained by using multiple sensors with different catalyst formulations enhancing rates species selectively. Some selectivity can also be obtained using diffusion constant differences among the species as discussed for the dual zirconia cell pumping devices. Selectivity considerations are already important in view of some on board diagnostic requirements in present legislation, where the monitoring of the three way catalyst is specified in terms of its efficiency for oxidation of non methane hydrocarbons (23,24). This issue of species specific response might become even more critical in view of contemplated future emission limits being set in term of atmospheric reactivity of the exhaust, as different classes of hydrocarbons have substantially different reactivities (24). The reactivities considered in the regulations however are those involved in the tropospheric ozone generation, and do not correspond to catalyzed oxidation reaction rates in the sensors. The more selective the sensor, the more sensitive it has to be. Thus, selectivity and sensitivity development for these sensors have to be high priorities in order for them to remain useful in the technologies needed to comply with these additional emission regulations. Along these lines, breakthroughs in spectroscopic sensing device development might offer a competitive technology in a somewhat longer time scale. The use of these devices as combustible detectors in engine operation feedback control, especially under conditions of unstable combustion, such as cold start, will require sensitivity and stability improvements. As difficult as these might be, the effort in realizing them is made worth while as uncontrolled cold start is responsible for a very large fraction of the automotive emissions.

Acknowledgments We wish to thank Drs. A. Lloyd Spetz, A. Baranzahi and Prof. I. Lundström who helped to get us going with the SiC MOS sensors and provided us with them. We are also grateful to the very helpful discussions with colleagues, especially M. Zanini-Fisher, R. E. Soltis and W. E. Weber of this laboratory.

Literature Cited

1.   Woestman, J. T.; Logathetis, E. M. *The Industrial Physicist* **1995**, *1*, pp 20-23
2.   Brailsford, A. D.; Yussouff, M.; Logothetis, E.M. *Sens. and Actuators* **1993**, *B13-14*, pp 135-138
3.   Logothetis, E. M.; Hetrick, R. E. In *Fundamentals and Applications of Chemical Sensors;* Schuetzle, D.; Hammerle, R. H., Eds, ACS Symposium Series No. 309; American Chemical Society: Washington DC, **1986**; pp 136-154
4.   Logothetis, E. M. *Ceramic Engineering and Science Proceedings* **1987**, *8*, pp 1058-1073
5. Logothetis, E. M., Visser, J. H., Soltis, R. E. and Rimai, L. *Sens. and Actuators B9* (1992) 183-189
6.   Visser, J. H.; Rimai, L.; Soltis, R.E.; Logothetis, E.M. *Digest of Technical Papers, The 7th International Conference on Solid State Sensors and Actuators (Transducers 93)*, Yokohama, **1993**; pp 346-349
7.   Visser, J. H.; Logothetis, E. M.; Rimai, L.; Soltis, R. E. US Patent No 5,281,313 Jan. 25, 1994
8.   Visser, J. H.; Soltis, R. E.; Rimai, L.; Logothetis, E. M. *Sens. and Actuators* **1992**, *B9,* pp 233 - 239
9.   Nicollean, E. H.; Brews, J. R. *MOS Physics and Technology*; John Wiley and Sons, New York, **1982**; pp 176-230
10.   Visser, J. H.; Zanini, M.; Rimai, L.; Soltis, R. E.; Kovalchuk, A.; Hoffman, D. W.; Logothetis, E. M.; Bonne, U.; Brewer, L. T.; Bynum, O. W.; Richard, M. A. *Technical Digest, 5th International Meeting on Chemical Sensors,* Rome, **1994;** pp 468-471
11.   Jones, E. *Solid State Gas Sensors*, Moseley, P. T.; Tofield, B. C. Eds, A. Hilger, Bristol, UK, **1987;** pp 17-31
12.   Zanini, M.; Visser, J. H.; Rimai, L.; Soltis, R. E.; Kovalchuk, A.; Hoffman, D. W.; Logothetis, E. M.; Bonne, U.; Brewer, L. T.; Bynum, O. W.; Richard, M. A. *Sens. and Actuators* **1995**, *A 48,* pp 187 - 192
13.   Lundstrom, I.; Shivaraman, M. S.; Svensson, C. *J. Appl. Phys.* **1975**, *46*, pp. 3976-3879
14.   Svensson, C. M., Lundkvist, L. S. and Lundström, I US Patent No. 04058368, November 1977
15.   Poteat, T. L.; Lalevic, B.; Kuliyev, B.; Yousuf, M.; Chen, M. *J. Electronic Materials* **1983**, *12*, pp. 181-214
16.   Lundström, I. *Sens. and Actuators* **1981**, *1*, pp. 403-426
17.   Lundströom, I., Armgarth, M. and Petersson, L.-G., *CRC Critical Reviews in Solid State and Material Science* **1989**, *15*, pp 201-277
18.   Baranzahi, A. *High Temperature Solid State Gas Sensors Based on Silicon Carbide*; Linköping Studies in Science and Technology 422; Linköping, **1995**.

19. Arbab, A.; Spetz, A.; Lundström, I. *Sens and Actuators* **1993**, *B15-16*, pp 19-23

20. Baranzahi, A., Spetz, A. L., Andersson, B. and Lundström, I. *Sens. and Actuators* **1995**, *B26-27,* pp. 165-169

21. Hunter, G. W., Neudeck, P. G., Chen, L-Y., Knight, D., Liu, C.C. and Wu, Q. H. *Silicon Carbide and Related Materials 1995, Proc. of the 6th Int. Conf.* Kyoto, 1995, Inst. of Physics Conference Series No. 142; Kyoto, **1995,** pp 817-820

22. Baranzahi, A., Spetz, A. L., Glavmo, M., Nytomt, J. and Lundström, I. *Transducers 95 and Eurosensors IX*, Vol. 1, Stockholm **1995,** pp 741-744

23. Kaiser, E. W., Siegl, W. O., Henig, Y. I., Anderson, R. W. and Trinker, F. H. *Env. Sci. Technol.* **1991**, *25*, pp. 2005-2011

24. Kaiser, E. W.; Siegl, W. O.; Anderson, R. W., *Fuels and Lubricants Meeting,* , SAE Paper No. 941960, Baltimore October **1994**.

Chapter 10

Silicon-Based Sensor for Flourine Gas

W. Moritz[1], S. Krause[1,3], Lars Bartholomäus[1], Tigran Gabusjan[1], A. A. Vasiliev[2], D. Yu. Godovski[2], and V. V. Malyshev[2]

[1]Humboldt-University Berlin, Walther-Nernst-Institute of Physical and Theoretical Chemistry, Bunsenstrasse 1, 10117 Berlin, Germany
[2]RRC "Kurchatov Institute", 123182 Moscow, Russia

A new Sensor for the determination of fluorine has been developed using a simple structure $Si/LaF_3/Pt$. A thin layer of the ionic conductor LaF_3 is in direct contact to the semiconductor without intermediate insulator. The field effect in the semiconductor leads to much steeper capacitance-voltage curves than usually found for conventional MIS-structures. It was shown that a special treatment of the sensor is necessary before first use to achieve a stable response. Different ionic conductors and gate materials were tested and the preparation conditions varied. The lower limit of detection of the optimised sensor was shown to be < 0.1 ppm. It was proven that potential formation takes place at the three phase boundary $LaF_3/Pt/gas$. Oxygen can be determined with a similar sensor system but does not interfere for the fluorine sensor.

Fluorine is used for several industrial processes for example in the production of polymers or the preparation of nuclear fuels. On the other hand, the gas is toxic even at low concentrations. Some Fluorohydrocarbons are known to cause critical changes in the ozone concentration in the atmosphere. Consequently, substitutes are developed as for example R 134a. Therefore, there is a great demand for sensors for these gases and alarm systems controlling the environment.

Murin et al. (*1*) have proposed to use a single crystal of LaF_3 covered with a platinum film to develop a potentiometric device for the determination of fluorine. Drawbacks of this sensor are the high price of the single crystal and the formation and stability of a reversible back side contact.

Low temperature oxygen sensors using the LaF_3/Pt contact have also been developed (*2-5*). Previously, we have proposed a silicon based sensor (*3*). The consecutive deposition of thin films of LaF_3 and Pt on silicon wafers results in a

---

[3]Current address: Department of Chemistry, University of Sheffield, Sheffield S10 2TN, England.

©1998 American Chemical Society

field-effect semiconductor sensor. We have shown that for these devices a stable sensor signal can be obtained without a reversible back side contact of the LaF_3. The extremely high steepness of the capacitance-voltage curves caused by the direct contact of the silicon to the ionic conductor improves the accuracy of the detection of concentrations (6,7). The mechanism of the oxygen sensor has been shown to depend on the surface treatment of the active layers (8), which have to form a three phase boundary ionic conductor/platinum/gas (9).

The long-term stability of the oxygen sensor has been improved by a simple thermal activation process (10). Using this activation the increase of the response time with the measuring time is not only stopped but even a faster response is obtained.

In this work we used the two different states of the sensor (reactivated or aged for some month after preparation) in combination with specific treatments of the surface to investigate the sensitivity to gases containing fluorine or flourohydrocarbons. An improvement of the electrical characteristics of the sensor was achieved by doping the LaF_3-layer with SrF_2.

Experimental

The sensor used for measurements in gases was a simple thin layer structure on a silicon substrate as described in (11). The LaF_3 layer (d=250nm) was prepared by thermal vapour deposition. Platinum (d=30nm) was deposited by sputtering in an Argon plasma so that a three phase boundary LaF_3/Pt/gas was created. Only minor variations in the preparation conditions were necessary compared to the oxygen sensor .

The sensitivity of the sensor was determined using high-frequency capacitance-voltage (CV) measurements or photo-current/voltage curves. Gas concentrations were adjusted using flow controllers controlled by a PC. All measurements for fluorine gas were performed at room temperature.

The sensors were activated on a hot plate at a temperature of 350°C for 2 minutes or by electrical current as described below.

Results

Electrical Properties of the Metal/Ionic Conductor/Semiconductor Sensor.
Chemical sensors based on the field effect in a semiconductor substrate can be characterised as transistor devices or using high-frequency capacitance voltage curves of simple thin layer structures. The shift of the CV-curve on the voltage axis with changing concentration of the analyte represents the voltage drop at the potential determining interface. Normally, the gate area of chemically sensitive field effect devices is covered by insulators such as SiO_2, Si_3N_4, Al_2O_3 or Ta_2O_5. While these layers are directly sensitive to the pH of a solution, additional layers on top of the insulator result in a sensitivity to other species. We have shown (6,7) that the direct contact of the semiconductor to the ionic conductor LaF_3 (without insulating interlayer) leads to a field effect device too and results in a considerable improvement of the capacitance-voltage curves (compare Figure 1 curves a and b).

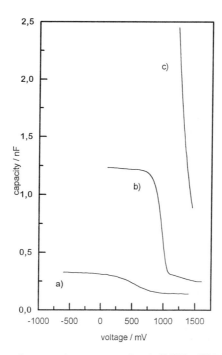

Fig.1 Capacitance voltage curves for a) Si/SiO$_2$/Si$_3$N$_4$/LaF$_3$/Pt;
b) Si/LaF$_3$/Pt and c) Si/LaF$_3$(SrF$_2$)/Pt

To further improve the electrical properties of the structure, LaF_3 and SrF_2 were evaporated simultaneously. The doping of the thin LaF_3 layer with 0,4% SrF_2 leads to another significant increase in the capacitance and steepness of the CV-curve of the sensor. As shown in Figures 1 and 2 a factor of more than 20 was achieved for the capacitance in the accumulation region.

Impedance spectra showed that the ionic conductivity in the LaF_3 layer doped with SrF_2 is by two orders of magnitude higher compared to the pure LaF_3, i.e. for frequencies from 100 Hz to 100 kHz the impedance of the doped layer is not determined by the bulk capacitance anymore because the capacitive impedance is smaller than the ohmic impedance in parallel. Therefore, the capacitance of the sensor in accumulation and its electrical behaviour is determined by the capacities of the interface LaF_3/Pt and LaF_3/Si. For these capacities values of 0,95 mF/cm^2 were found which corresponds to a distribution of F⁻-ions in a thin space charge region.

Electrical Activation of the Sensor. Using the Si/LaF_3/Pt sensor structure for oxygen measurements we obtained a stable sensitivity of 58 $mV/\Delta lg\ p_{O_2}$ and a response time of t_{90}=70s. The flat band voltage and the sensitivity were highly reproducible. However, a drawback for the application of the sensor were a decrease of the response rate as a function of time (10). While for several days after the preparation of the platinum layer reasonable times are obtained, the sensor response gets too slow for an application after a period of 2 weeks. The comparison of different methods of reactivation has shown that a thermal treatment at 350°C in air is the most promising. Using this treatment the sensor is slightly faster than a freshly prepared one and the sensor drift is reduced to 0.5 mV/h. The heating procedure can be repeated any number of times. Heating the sensor once a week, a stable response time was obtained for a period of 4 month. Even 8 month after preparation a full activation was observed. It is noteworthy that the behaviour of the reactivated sensor is similar to that of a newly prepared one but there are some differences concerning the drift and the influence of humidity on the response time (11).

External heating of the sensor would be a disadvantage for sensor application. The integration of a heater is one way to a low-power on-chip reactivation. Since the sensor reaction takes place at the three phase boundary LaF_3/Pt/gas, only the thin platinum layer has to be be heated. This could be achieved using the platinum layer not only as an electrode but also as an electrical heater. The principle of sensor operation and activation is shown in Figure 3. The characterisation of the structure with photocurrent measurements was more convenient for these experiments because the active area was determined by the light and there was no influence of the contact areas.

It was shown that the duration of the electrical heating can be reduced to very short times. Even an impulse of 100 ms gives a full reactivation of the sensor. For such short times the heat transport into the silicon becomes negligible. Hence, the temperature of the silicon chip is increased by only 0.4 °C and a measurement at room temperature is possible directly after the thermal treatment.

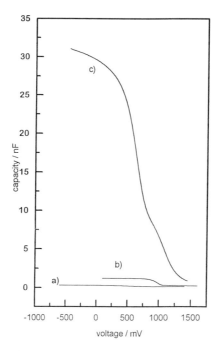

Fig.2 Capacitance voltage curves (Scale changed) for
a) $Si/SiO_2/Si_3N_4/LaF_3/Pt$; b) $Si/LaF_3/Pt$ and c) $Si/LaF_3(SrF_2)/Pt$

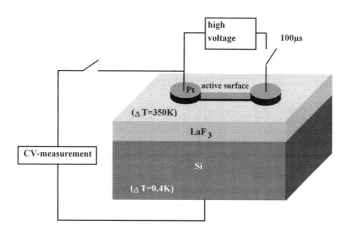

Fig.3 Scheme of sensor operation and activation

Sensitivity to Fluorine. The oxygen sensitive structure $Si/ LaF_3/ Pt$ could also be used as a fluorine sensor, but an optimisation of the preparation conditions for this application was necessary. Furthermore the influence of the thermal activation process was investigated.

For a non-activated sensor aged for several month neither sensitivity for oxygen nor for fluorine was found. Therefore the thermal activation of the sensor is the key process to obtain a fluorine sensitivity.

After the thermal activation of the sensor the potential was shifted for about + 800 mV in fluorine containing gases compared to measurements in oxygen/ inert gas mixtures. The response of the sensor to different concentrations of fluorine in synthetic air at room temperature is shown in Figure 4. A constant potential was reached within a few minutes after the concentration change even for low concentrations. Sensitivity measurements showed a linear relationship between the potential and the logarithm of the fluorine partial pressure. This corresponds to the Nernst equation, but sensitivities were found to be $111(\pm 11)mV/lg\ p(F_2)$ (Figure 5) corresponding to a formal exchange of about 0.5 electron. Furthermore, it was observed, that the response time increases with decreasing fluorine concentration. The lower limit of detection was shown to be 0.1 ppm. In contrast to the behaviour of the oxygen sensor no decrease in response rate with time was found for the measurements in fluorine gas.

For the understanding of the mechanism of the sensor it is important to know that the shift in potential by about +800 mV after exposure to a fluorine containing gas is a slow process. As shown in Figure 6 the potential was increasing with the time of preconditioning and the concentration of fluorine. There were strong improvements in sensitivity, response time and a reduction of drift after a pretreatment in 1000 ppm fluorine. This behaviour is an argument for the existence of a mixed potential.

A structure $Si/SiO_2/Si_3N_4/Pt$ was tested for fluorine sensitivity too. As there was no sensitivity found, the existence of the three phase boundary fluoride ionic conductor/platinum/gas was proven to be necessary. On the other hand we could substitute the LaF_3 by BaF_2 getting the same results in sensitivity and response time.

Furthermore, for sensors exposed to fluorine no oxygen sensitivity was found in gases containing flourine or fluorine free gas mixtures.

Sensitivity to Fluorohydrocarbons. Beside the sensitivity to fluorine the possibility of a detection of HF in gases was shown (*11*). Therefore, we have been interested in the possibility of the specific detection of fluorohydrocarbons using our sensor.

At temperatures above 650K the equilibrium between 1,1,2,2- and 1,2,2,2-tetrafluorethane (R134)

$$CHF_2\text{-}CHF_2 \quad \overset{HF}{\rightleftarrows} \quad CH_2F\text{-}CF_3 \qquad (1)$$

can be obtained at catalysts. This should result in the intermediate existence of HF.

As shown in Figure 7 the detection of R134a was possible at temperatures as low as 155 to 180°C using the $Si/LaF_3/Pt$ sensor structure.

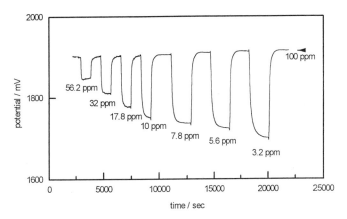

Fig.4   Sensor response to different concentrations of fluorine

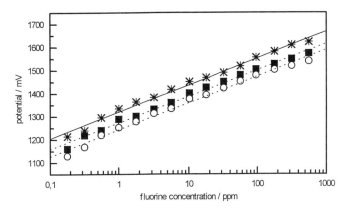

Fig.5   Sensitivity of the Si/LaF$_3$/Pt sensor structure to fluorine

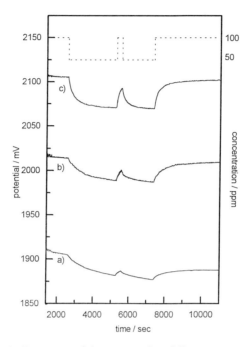

Fig.6   Response of the sensor after different treatment
a) 5minutes 100 ppm fluorine
b) 5 minutes 1000 ppm flurine
c) 2 hours 1000 ppm fluorine

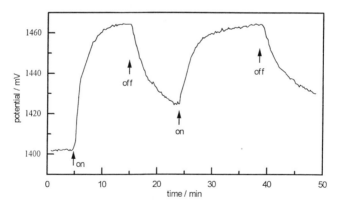

Fig.7   Response to 1,2,2,2-tetrafluorethane (R134a) in synthetic air at 180°C

Discussion

The sensitivity to fluorine could be explained by the formation of F⁻ according to

$$F_2 \quad + \quad 2e- \quad \rightleftarrows \quad 2F^- \qquad (2)$$

But this would result in a sensitivity of only 30 mV/decade. The sensitivity of the similar structure to oxygen and the very high sensitivities lead to the assumption that there might be some kind of mixed potential including a reaction with oxygen. The potential difference of about 2 Volts between the standard potentials of the fluorine and the oxygen reaction in aqueous solutions is an additional argument for this assumption. In (10) the rate determining process for the reaction at the oxygen sensor was shown to be:

$$O_{2ads} \quad + \quad H_2O \quad + \quad e- \quad \rightleftarrows \quad HO_{2ads} \quad + \quad OH^- \quad (3)$$

It has to be considered that there is not a real equilibrium, but the rates of the electrochemical reactions (2) and (3) determine the mixed potential. The existence of different oxygen species at the LaF_3 surface has been proven using XPS.

Specific detection of fluorohydrocarbons can be achieved assuming that the decomposition of the organic compound is followed by recognition of fluorine. It is noteworthy that the detection of the R134a was possible at a temperature 200°C lower than the temperature at which the best catalysts known for the decomposition of this substance are working. This is because the decomposition and the detection occur at the same place at the three phase boundary LaF_3/Pt/gas.

All the results given here were obtained from measurements with capacitive devices but field effect transistors can be produced in a similar way.

The use of two sensors, one in the aged state and the other activated by an electrical impulse results in a sensor system for the simultaneous determination of fluorine, hydrogen fluoride and fluorohydrocarbons. Furthermore, it is the same sensor that can be used for the determination of oxygen. Since there is no demand to determine all the gases at the same time the importance of selectivity is reduced. A sufficient selectivity for these gases is achieved by the difference in conditions of measurements as summarised in Table I.

128

Table I Differences in conditions for the determination of various gases using the Si/LaF$_3$/Pt sensor structure

Sensor / Conditions	Oxygen	Hydrogen-fluoride	Fluorine	C$_2$H$_2$F$_4$
activated	x	-	x	x
non activated	-	x	-	-
Voltage range[V]	1,2-1,5	1,2-1,5	2,0-2,5	1,0-1,5
Temperature	room temperature	room temperature	room temperature	440 K

Literature Cited

1   I.W. Murin, O.W. Glumov, D.B. Samusik; Potentiometric sensor for the determination of fluorine in gases, Zh. Prikl. Khim., 64 (10) **1991** 2171
2   N. Miura, J. Hisamoto and N. Yamazoe, Solid state oxygen sensor using sputtered   LaF$_3$ films, Sensors and Actuators, 16 **1989** 301-310
3   S. Krause , W. Moritz, I. Grohmann, A low-temperature oxygen sensor based on the Si/LaF$_3$/Pt capacitive structure, Sensors and Actuators B, 9 **1992**, 191
4   J.P. Lukaszewitz, N. Miura and N. Yamazoe, Influence of water treatment of LaF$_3$ crystal on LaF$_3$-based oxygen sensor workable at room temperature, Jpn. J. Appl. Pys., 30 **1991** L1327
5   S. Harke, H.-D. Wiemhöfer and W. Göpel, Investigations of electrodes for oxygen sensors based on lanthanum trifluoride as solid electrolyte, Sensors and Actuators, B1 **1990** 188-194
6   J. Szeponik, W. Moritz, A New Structure for Chemical Sensor Devices, Sensors and Actuators, 2 **1990** 243
7   J. Szeponik, W. Moritz, F. Sellam, LaF$_3$ Thin Films as Chemical Sensitive Material for Semiconductor Sensors, Ber. Bunsenges. Phys. Chem. 95 **1991** 1448
8   S. Krause, R. Krankenhagen, W. Moritz, I. Grohmann, W. Unger, Th. Gross, A. Lippitz, Influence of the LaF$_3$/metal interface on the properties of

a low temperature oxygen sensor, Eurosensors VI, San Sebastian, Spain, 5.-7.10.1992 and Sensors and Actuators, B16 **1993** 252

9   S. Krause, W. Moritz, I. Grohmann, W. Unger, Th. Gross, A. Lippitz, Dynamic response of a low-temperature field effect oxygen sensor, International Meeting on Chemical Sensors, Tokyo, Japan, 13.-17.9.1992 and Sensors and Actuators B, 14 **1993** 499

10   S. Krause, W. Moritz, I. Grohmann, Improved long-term stability for a LaF_3 based oxygen sensor, Proc. EUROSENSORS VII, 1993 Budapest, Hungary and Sens. and Act., B 18 **1993** 148-154

11   W. Moritz, S. Krause, A.A. Vasiliev, D.Yu. Godovski and V.V. Malyshev, Monitoring of HF and F_2 using a field effect sensor, The Fifth International Meeting on Chemical Sensors, Rome 11-14. July 1994 and Sens. and Act. B, 24-25 **1995** 194

Chapter 11

Cyclodextrin-Based Microsensors for Volatile Organic Compounds

B. Swanson, S. Johnson, J. Shi, and X. Yang

Chemical Science and Technology, Los Alamos National Laboratory,
Los Alamos, NM 87545

Guest-host chemistry and self-assembly techniques are being explored
to develop species selective thin-films for real-time sensing of volatile
organic compounds (VOCs). Cyclodextrin (CD) and calixarene (CA)
molecules are known to form guest-host inclusion complexes with a
variety of organic molecules. Through the control of the cavity size
and chemical functionality on the rims of these bucket-like molecules,
the binding affinities for formation of inclusion complexes can be
controlled and optimized for specific molecules. Self-assembly
techniques are used to covalently bond these reagent molecules to the
surface of acoustic transducers to create dense, highly oriented, and
stable thin films. Self-assembly techniques have also been used to
fabricate multilayer thin film containing host reagents through
alternating adsorption of charged species in aqueous solutions. Self-
assembly of polymeric molecules on the SAW device was also
explored for fabricating species selective thin films.

The quest for chemical and biochemical microsensors has focused the need to create
highly selective and highly sensitive interfaces. Various chemical species such as
metals (1), metal oxides (2), metal complexes (3), organic host molecules (4), and
functionalized polymers (5) have been deposited on the transducer surface and
studied as molecular recognition reagents for sensing chemical species in the gas phase.
Moreover, organized self-assembly monolayers with functional groups such as
carboxylic acid, methyl and metal ions were examined as chemically selective
interfaces for sensing amines and VOCs (6). Among them, organic host molecules
containing a hydrophobic cavity attract increasing attention in sensor development for
real-time monitoring of volatile organic compounds (VOCs). These organic receptors
bind VOCs through noncovalent interactions such as van der Waal's and hydrogen
bonding, relying heavily on their hydrophobic cavity.

©1998 American Chemical Society

In our research on VOC sensor development, we hope to achieve highly selective and sensitive sensors based on guest-host chemistry. By changing the functionality and the cavity sizes of CDs, chemical specificity of guest-host interactions can be achieved. Three approaches were studied to incorporate cyclodextrin (CD) molecular recognition reagents into interfaces on surface acoustic wave (SAW) devices (7). The first one relies on molecular self-assembly monolayer techniques to couple CDs covalently to the transducer surface as densely packed and highly orientated monolayers. The second approach employs a polyelectrolyte self-assembly technique to fabricate organized multilayer interfaces containing CDs and calix(n)arenes (CAs). The third approach explores *in-situ* polymerization reactions to cross-link derivatized CD molecules onto the transducer surface.

Molecular recognition reagents. Cyclodextrins (CDs) are natural products from starch and are produced on the industrial scales. CDs have a preorganized rigid cavity with three different sizes and the hydroxyl groups on the rims of the cavity allow for various chemical transformations (Figure 1). CDs are best known to form guest-host complexes in solutions, especially in aqueous media, with organic species (8). However, derivatized cyclodextrins immobilized on capillary columns have been used for separations of a variety of organic species in gas phases at elevated temperatures (>50 ^{0}C) (9). The fact that organic enantiomers can be resolved by CD coated columns confirms that CDs bind the gas species in their cavities (9a). A recent

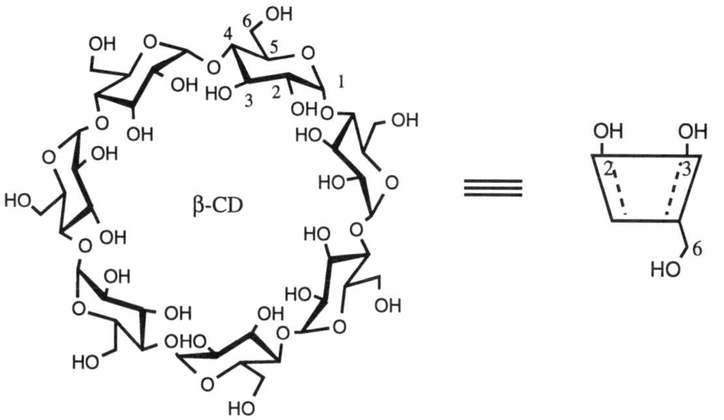

Figure 1. Molecular structure of β-cyclodextrin and schematic representation of the cyclodextrin shape and secondary hydroxyl groups on upper rim and primary hydroxyl groups on the lower rim.

theoretical calculation on the permethylated β-CD revealed that the guest-host interaction is mainly controlled by van der Waals forces (10). In the gas phase the host-guest electronic complementarity, steric complementarity (shape recognition) and the host preorganization are three key elements in determining the stability constant of a supramolecular complex (11).

Functionalized cyclodextrins have been studied for various applications such as enzyme mimics, separations, and drug delivery (12). The binding characteristics of CDs change upon functionalization of the hydroxyl groups, because of changes in the shape, size and charge distributions of the cavity and modification of intramolecular hydrogen bonding (13). By creating a series of alkylated cyclodextrins and leaving the primary hydroxyl groups unchanged for future coupling to the transducer surface, we were able to study the different binding affinities towards VOCs based on the changes in the R group (Figure 2) and utilize these differences to create chemically specific sensors.

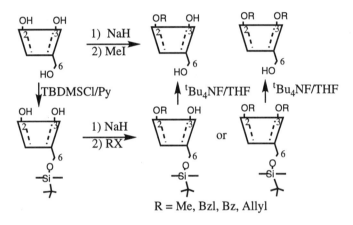

R = Me, Bzl, Bz, Allyl

Figure 2. Chemical transformations of cyclodextrins (14).

Self-assembled monolayers as sensing interfaces. SAMs or LB films have been used as a linker layer to immobilize enzymes and antibodies on biosensors (15). Similarly, we explored the possibility to use SAMs as linker layers to covalently couple cyclodextrin reagents to the transducer surface, in this case, surface acoustic wave device. It has been recognized that silanes with long alkyl chains form SAMs with organized quasi-crystalline structures. A bromo-terminated SAM was formed by the immersion of a quartz substrate into a solution of 11-bromo-undecyltrichlorosilane or 16-bromo-hexadecyltrichlorosilane hexadecane solution. The nucleophilic substitution reaction on the SAM with potassium cyanate gave an isocycanate-terminated surface, which reacts in situ with primary hydroxyl groups on the CDs (Figure 3). The monolayers were characterized by ellipsometry measurement, contact angle measurement and FT-IR.

Several SAW devices coated with monolayer cyclodextrins were tested for sensing volatile organic compounds. Figure 4 is the real time responses of a heptakis(2-O-benzyl)-β-cyclodextrin SAW device to halogenated hydrocarbons. The

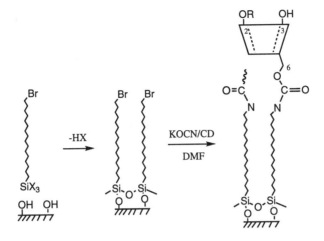

Figure 3. Formation of bromo-terminated SAM and coupling of cyclodextrins to the transducer surface.

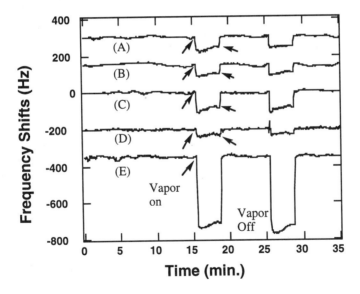

Figure 4. Real time SAW responses to halogenated hydrocarbons: (A) 65 Hz at 1115 ppm 1,1,1-trichloroethane; (B) 58 Hz at 552 ppm trichloroethylene; (C) 103 Hz at 1644 ppm chloroform; (D) 38 Hz at 138 ppm perchloroethylene; (E) 377 Hz at 1381 ppm perchloroethylene. The device was coated with heptakis(2-O-benzyl-β-cyclodextrin).

sensor response is reversible, rapid and linear, as shown in the case of perchloro-ethylene (Figure 4(D) and 4(E)). The sensor is much more sensitive toward perchloroethylene and toluene (not shown here) and less sensitive to chloroform and 1,1,1-trichloroethane which are relatively larger in size. We also found that the sensitivity of the device is much lower towards smaller molecules such as methanol, hexane and acetone. In general, monolayer CD coated SAW devices can detect VOC vapors at about 100 ppm concentration.

Polyelectrolyte deposition approach to mutilayer CD and CA films. For developing selective sensor coatings containing molecular recognition reagents such as cyclodextrins, calixarenes, cavitands, metalloporphyrins, there exist problems in fabricating ordered multilayer molecular films. Solution casting of molecular species results in films with low coverage, poor uniformity, low stability and low reproducibility. We studied a recently developed technique using polyelectrolyte self-assembly to grow organic mutilayer films. This technique consists of alternating adsorption of anionic and cationic electrolytes on a charge substrate by sequential dipping of the substrate into aqueous polyelectrolyte solutions. The films formed are uniform, organized, durable and reproducible in terms of film thickness (16). Multilayer thin films with precisely controlled thickness and molecular architectures can therefore be successfully fabricated. Replacing polyanions, highly negatively charged molecular recognition reagents were employed for film fabrication. This unique approach incorporates polymer and molecular elements into the sensing film and thus results in films with polymer's physical properties and molecular film's selectivity. Multilayer molecular films of polyelectrolyte/calixarene and polyelectro-lyte/cyclo-dextrin hosts were fabricated by alternating adsorption of charged species in aqueous solutions onto a substrate (quartz or silicon wafer). 250 MHz SAW device was first treated with aminopropyltrimethoxysilane, followed with deposition of poly(sodium 4-styrenesulfonate) (PSS) and then poly(dimethyldiallyl-ammonium) (PDDA). After this, alternating depositions of negatively charged molecular host species (Figure 5) and PDDA were carried out until the desired number of bilayers was reached. Between each deposition, the device was thoroughly rinsed with deionized water. For film characterizations, silicon wafers or quartz substrates were used for film depositions. This layer-by-layer molecular deposition approach has been successfully employed to integrate molecular recognition reagents into the polymer films as chemically selective layers. The films were characterized with SEM, infrared and mass loading was monitored by SAW devices. These measurements revealed that the deposition process is highly reproducible and the resulting films are uniform and stable.

SAW sensors coated with the calix[6]arene sulfonate/PDDA (Figure 5) or sulphated β-cyclodextrin/PDDA were studied for sensing a variety of organic solvent vapors. It was found that sensor coated with the calixarene showed increased sensitivity with the increasing number of calixarene layers deposited on the SAW device. However, the sensitivity to carbon tetrachloride and 1,1,1-trichloroethane increases only slightly. The sensor responses to perchloroethylene (PCE) increase

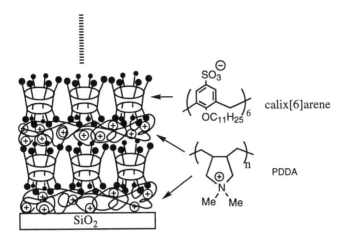

Figure 5. Alternating polyelectrolyte PDDA and p-sulfonated calix[6]arene depositions on a surface-modified SiO₂ surface.

with increasing number of the calix[6]arene layers. The result indicates that , in addition to the surface layer, the vapor sorption also involves inner layer species. The five bilayer calix[6]arene film exhibts high sensitivity towards PCE, toluene and trichloroethylene (Figure 6).   The sensor can detect vapors at sub-ppm concentrations.   In contrast, a SAW device coated with only polyelectrolytes (PDDA/PSS) didn't show increased sensitivity upon increasing thickness of the film

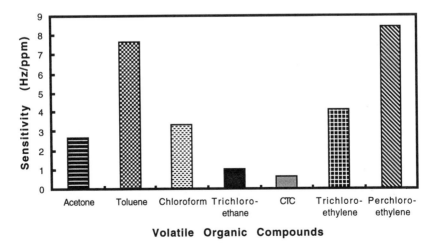

Figure 6. Sensor sensitivity to volatile organic compounds. The sensor was coated with five bilayer of p-sulphonated calix[6]arene/PDDA.

and the sensitivity is much lower (3 to 4 times less). There is no selectivity observed with this film. We also fabricated a sulphated β-cyclodextrin bilayer film with PDDA. The sensor was not particularly sensitive towards hydrophobic organic vapors, which might result from high charge density and hydrophilic nature on the CD rim. Other charged CD species are now being pursued.

In-situ formation of polysiloxane cyclodextrin films on the surface. This is an in-situ polymerization approach to multilayer selective and sensitive sensing layers. The purpose of using a siloxane formation approach is two-fold. First, this provides a simple and easily reproducible approach to multilayer films for optimization of sensitivity. Secondly, the use of a siloxane intervoid network to minimize adsorption of target VOCs into interstitial sites that will act as competitive binding sites relative to the cyclodextrin cavities. The hydrophobic VOCs would prefer to be adsorbed in the molecular host cavities. In general, SAW devices were coated by immersing the devices in a DMF solution of trimethoxysilane functionized CDs overnight (Figure 7). The devices was then cured in the air at 60 °C for 30 min. The mass loading of the polymers on SAW devices ranges from 128 to 138 kHz, corresponding to the film thickness of 11 to 13 nm. The films are uniform and amorphous, as revealed by SEM.

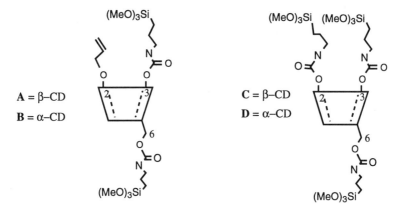

Figure 7. Siloxane CD polymer precursors.

The siloxane CD coated devices are sensitive and selective toward several VOCs. The response time is in seconds of the exposure to the analytes and the responses are reversible. CD-siloxane coated SAW sensors can detect analytes (perchroloethylene, trichloroethylene, trichloroethane, etc.) in tens of ppm without difficulty. We also tested the stability of these CD-siloxane coated devices. After 2 months, no significant sensitivity loss was observed. Figure 8 shows response patterns for eight vapors over four CD-siloxane coated SAW sensors. These eight vapors can be grouped into three, according to their response patterns. As expected, they are grouped by their chemical structures: nonpolar hydrocarbons (hexane and toluene), polar hydrocarbons (acetone and methanol), and halogenated hydrocarbons

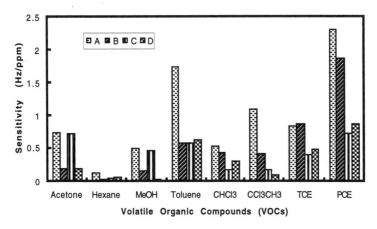

Figure 8. Sensor array response patterns to VOCs with four 250 MHz SAW devices coated with siloxane CD polymer precursors A, B, C, and D.

(chloroform, trichloroethane, and perchloroethylene and TCE).   Hexane has a very low sensitivity, compared with toluene. The four halogenated hydrocarbons, though similar, have distinct patterns.   An α-CD coated SAW sensor is more selective to toluene and hexane than β-CD coated sensors.   On the other hand, a β-CD coated sensor is more selective to halogenated hydrocarbons. The resulting selectivity due to CD cavity size difference indicates that vapors do recognize their hosts at the gas-solid interface. The affinity of toluene toward α-CD and PCE towards β-CD is well documented in the solution studies and solid state structures.

Conclusion

Three approaches have been studied to fabricate chemically selective interfaces for sensing volatile organic compounds. By incorporating molecular recognition reagents, cyclodextrins and their derivatives, into sensing layers, SAW sensors with high sensitivity and selectivity can be obtained.   Further work includes tailoring of cyclodextrins and other cavity containing molecular reagents for the detection of chemical warfare agent and explosive vapors as well as the construction of organized sensing mutilayers and minimizing non-specific binding.

Acknowledgments

This work was supported by Laboratory Directed Research and Development (LDRD) funds and the Advanced Technologies for Proliferation Detection program of the Nuclear Nonproliferation Office of the Department of Energy.

138

References

1. D'Amico, A.; Palma, A.; Vetelino, E. *Sens. Actuators* **1982**, *3*, 31.
2. Bryant, A.; Poirier, M.; Riley, G.; Lee, D. L.; Vetelino, J. F. *Sens. Actuators* **1983**, *4*, 105.
3. (a) Zeller, E. T.; Zhang, G. *Anal. Chem.* **1992**, 64, 1277; (b) Nieuwenhuizen, M. S.; Harteveld, J. L. *Talanta* **1994**, *41*, 461.
4. (a) Lai, C. S.; Moody, G. J. Thomas, J. D. R *J. Chem. Soc. Perkin Trans.* **1988**, 319; (b) Bodenhöfer, K.; Hierlemann, A.; Noetzel, G.; Weimar, U.; Göpel, W. *Anal. Chem.* **1996**, *68*, 2210; (c) Schierbaum, K. D.; Weiss, T.; Thoden, E. U.; Velzen, T.; Engbersen, J. F. J.; Reinhoudt, D. N.; Göpel, W. *Science* **1994**, *265*, 1413; (d) Moore, L. W.; Springer, K.; Shi, J.; Yang, X.; Swanson, B. I.; Li, D. *Adv. Mater.* **1995**, *7*, 729;
5. McGill, R. A.; Abraham, M. H.; Grate, J. W. *Chemtech* **1994**, 27.
6. Thomas, R. C.; Yang, H. C.; DiRubio, C. R.; Ricco, A. J.; Crooks, R. M. *Langmuir* **1996**, *12*, 2239.
7. Wohltjen, H.; Dessy, R. E. *Anal. Chem.* **1979**, *51*, 1458;
8. Saenger, W. in *Inclusion Compounds;* Atwood, J. L.; Davis, J. E. D,; MacNicol, D. D., Ed.; Academic Press, London, 1984, Vol.2; pp 231-260.
9. (a) König, W. A.; Lutz, S.; Hagen, M.; Krebber, R.; Wenz, G.; Baldenius, K.; Ehlers, J.; Dieck, H. tom. *J. High Res. Chromatorg.* **1989**, *12*, 35; (b) Schurig, V.; Schmalzing, D.; Schleimer, M. *Angew. Chem. Ed. Int. Engl.* **1991**, *30*, 987.
10. Lipkowitz, K. B.; Pearl, G.; Coner, B.; Peterson, M. A. *J. Am. Chem. Soc.* **1997**, *119*, 600.
11. Diederich, F. *Cyclophanes* The Royal Society of Chemistry, London, **1991**.
12. Szejtli, J. in *Inclusion Compounds;* Atwood, J. L.; Davis, J. E. D,; MacNicol, D. D., Ed.; Academic Press, London, 1984, Vol. 3; pp 331-391.
13. Harata, K. in *Inclusion Compounds;* Atwood, J. L.; Davis, J. E. D,; MacNicol, D. D., Ed.; Academic Press, London, 1991, Vol.5; pp 311-344.
14. (a) Takeo, K.; Mitoh, H.; Uemura, K. *Carbohydrate Research* **1989**, *187*, 203; (b) Rong, D.; D'Souza, V. T. *Tetrahedron Lett.* **1990**, *31*, 4273.
15. Thompson, M.; Krull, U. J. *Anal. Chem.* **1991**, *63*, 393A.
16. (a) Decher, G.; Hong, J. D. *Ber. Bunsenges. Phys. Chem.* **1991**, *95*, 1430; (b) Decher, G.; Hong, J. D.; Schmitt, J. *Thin Solid Films* **1992**, *210/211*, 831; (c) Lvov, Y.; Decher, G.; Möhwald, H. *Langmuir* **1993**, *9*, 481.

Chapter 12

A New Methodology for Testing and Characterization of Sorption Materials in a Gas-Flow System Based on Piezoelectric Sensors

Lj. V. Rajakovic

Department of Analytical Chemistry, Faculty of Technology and
Metallurgy, University of Belgrade, Karnegijeva 4, POB 494,
11 000 Belgrade, Yogoslavia

Activated carbon cloth (ACC) is useful as a sorbent in processes for
the complex purification of air and the separation of toxic pollutants
from air. Characterization of all types of cloths in a flow system
during sorption of organic vapours from the air is very important in
designing this new class of carbon material. Of particular importance
is the determination of the capacity and breakthrough point for each
type material and each specific pollutant. In this work, an attempt was
made to create a new sensor system for the characterization of ACC
and for monitoring the concentration profile during sorption process
based on the application of a pair of piezoelectric quartz crystals
(PQCs). One of PQCs was used as the referent inlet sensor and the
other were used as the indicator outlet sensor. In this way the inlet
concentration of the air-pollutant was permanently controlled, the
decrease in the pollutant concentration due to sorption onto the ACC
was monitored, the breakthrough point was determined and the
increase in the concentration after the breakthrough point was also
monitored.
The described methodology provides sensor response curves that are
convenient for displaying and analyzing the behaviour of the ACC and
of the PQCs. Benzene, toluene and phenols were used for testing.

The purpose of this work was to find a new convenient sensor system for the
characterization and evaluation of a sorption material for the removal of toxic
compounds from air. Accordingly the main objects of the investigations were 1)
piezoelectric gas sensors, 2) activated carbon cloth and 3) various organic vapours.
The study was divided into a theoretical and an experimental part.
The main interest in the work was to develop a methodology that would provide
reproducible and reliable recordings of the piezoelectric sensor responses during a

dynamic separation process. A piezoelectric quartz crystal has been successfully used as a gas-phase microgravimetric sensor and as a selective chemical sensor where a selective binding species was incorporated into a film deposited on the device surface (1-7). The high sensitivity, raggedness and low cost of this device makes it convenient for use in analytical chemistry. A bulk acoustic wave sensor in the thickness shear mode (BAW/TSM) has been used extensively in the gas phase and much less in the liquid phase (8). King (2) introduced the piezoelectric chemical sensor and addressed almost every aspect of its use. Most of the subsequent work was aimed at finding appropriate coatings for the assay of the most common air pollutants.

Particular interest was devoted to the development and evaluation of sorption material based on activated carbon cloth. Various types of activated carbon cloth (ACC) both with and without impregnation, were used. The ACC was activated by introducing a specific chemical agent inside the cloth structure (9-11). In this way, the possibility existed of separating toxic air pollutants efficiently by their bonding to one or more active chemical agents in the cloth. The activated carbon material, impregnated in such a way, can be used for producing filters for purification of air or industrial gases, both in peacetime and in war conditions, as well as for manufacturing protective clothes and associated masks. Impregnation was done by using a specific compound such as Triton X-100, (TX-100), poly(vinyl pyrrolidone), (PVP); 4-amino-antipyrine, (4-AAP) and metal-organic compounds such as copper tartarate and silver citrate. Metallic ions have a catalytic function, while organic components increase the sorptive capacity of the cloth. The content of each chemical agent in the ACC is determined by a convenient analytical method such as atomic absorption flame spectrometry and ion chromatography (11).

Benzene, toluene and phenols were used as the air-pollutants. Benzene was used as a standard adsorbate for physical absorption control whereas toluene and phenols were used as industrial and environmental hazardous chemicals of interest.

Experimental Section

The basic materials in this research were microporous ACC with a specific surface of 1500 m^2 g^{-1}, made by the carbonization and activation of a standard production type viscous rayon, produced by Toyobo, Osaka (type 1) and AJMV, Kijev (type 2) . The tartarates and citrates of copper and silver, as impregnation reagents, were used. The concentration of solutions of these salts was 0.1 mol dm^{-3}, and they were made from Merck and Zorka, Šabac chemicals. For the specific reagents solution as TX-100, PVP and 4-AAP concentration of solutions were 0.1%. The experimental procedures for the impregnation and characterization of the carbon cloth by classical methods have been described elsewhere (9-11). Usually ACC after impregnation was not washed. In same experiments washing procedure was applied and these cases are particularly noted.

An apparatus used for generating test vapours, exposing the ACC (16 cm^2) as sorption material and the coated PQCs as sensors and recording the responses in real time was evaluated. In designing the apparatus, special care was taken to prevent any changes in the crystal resonant frequency that could be caused by factors other than the presence of the chemical stimulants. The experimental data were collected automatically and simultaneously stored in a computer memory and sent to a recorder. The experimental set-up with the piezoelectric sensor system used for testing and characterization of the ACC is shown in Figure1.

It consists of three basic parts: a dynamic apparatus for the continuous and simultaneous generation of clean carrier gas and a constant concentration of target vapor, a cell with the ACC and a sensor system for collecting, recording and displaying data. A continuous flow system was used. The coated 9 MHz, Al-plated PQCs were exposed in sequence to the pure air, the air with a sample and the pure air again. The concentration of any sample was tested by GC (5). An optimal flow rate ($17.5 - 75.0$ cm^3 min^{-1}) with respect to sensor response and recovery time was used. All the experiments were performed at room temperature (20-25°C).

Results and Discussion

First it is necessary to define some characteristic points that are basic for the evaluation of the ACC. Idealized response curves with characteristic points are presented in Figure 2.

Curve "B" presents the response of the referent piezoelectric sensor placed in the inlet stream of vapour, it corresponds with the inlet concentration. Curve "A" presents the response of the indicator piezoelectric sensor placed in the outlet stream of vapour, after the ACC cell, it corresponds with the outlet concentration of the vapour which had penetrated through the ACC. The main objectives of the analysis are to predict the total quantity of the gas stream at the breakpoints, T_I, T_{II} and the saturation or exhaustion point, T_S. The primary breakpoint, T_I, is the moment when the adsorption zone reaches the opposite end of the ACC when the vapour concentration in the outlet stream starts to rise rapidly. It is indicated by a rapid decrease of the crystal frequency, according to Sauerbrey -$\Delta F = k \ \Delta c$, (1). Second breakpoint, T_{II}, is the moment when the chemical agent incorporated in the ACC structure ends its bonding function during the sorption process. After the second breakpoint little additional adsorption occurs since the entire ACC is approaching equilibrium with the inlet gas stream. At the end a total exhaustion can be observed in the saturation point, T_S. The time of protection for each type of ACC, the time before the penetration of each pollutant through the ACC can be analyzed by breakthrough points, T_I, T_{II}, and by the total amount of vapour removed from the gas stream, m. The shaded area in Figure 2 represents the amount of pollutant removed, m (in mg), from the gas by the ACC between the inlet and the exhaustion point which can be expressed by:

$$m = \int_{T_I}^{T_S} (C_0 - C) \ dQ;$$

$$m = \int_{T_I}^{T_S} (F_0 - F) \ dQ;$$

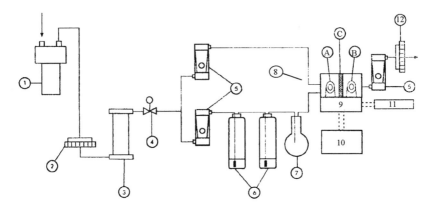

Figure 1. Experimental set-up with piezoelectric sensor system used in testing and characterization of activated carbon cloth.

1-air pump, 2-charcoal scrubber, 3-dryer column, 4-needle valve, 5-flow-meters, 6-bubblers, 7-dumper, 8-sensor cell (A-indicator sensor, B-referent sensor, C-ACC cell), 9-oscillator, 10-frequency counter, 11-power supply 9V, 12-charcoal filter.

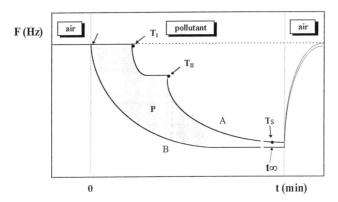

Figure 2. Typical idealized output of piezoelectric sensor system equipped with an indicator sensor (A) and a referent sensor (B). Note: breakthrough points T_I and T_{II} and saturation point T_S.

where: C_o is the inlet concentration of pollutant (in mg cm^{-3}) which corresponds with the resonant frequency F_o (in Hz), C is the concentration of pollutant in the gas stream in equilibrium which corresponds with the resonant frequency F, and Q is the flow rate (in cm^3 min^{-1}). The mass of vapour retained by the ACC can be analyzed through the saturation point (T_s), and calculated by applying a simple integration procedure.

In Figure 3 real recordings of data for benzene (C=319 mg dm^{-3}) passing through three characteristic ACCs (type 2) are presented: a) the ACC without impregnation, b) the ACC impregnated with copper tartarate and c) the ACC impregnated with Triton TX-100. In all the experiments the PQCs were coated with 10 µg of 4-amino-antipyrine, the flow-rate was 75 cm^3 min^{-1} and the temperature was 25° C.

These results confirmed that activated carbon cloth can be characterized by this simple procedure based on a pair of piezoelectric sensors. The time of protection until the penetration point (T_1) is the main criteria for this type of material and the answer can be obtain with this simple methodology. It was observed that the ACC without any additional impregnation was active and protective, benzene was retained-adsorbed 5 minutes. The inlet concentration of benzene of 319 mg dm^{-3} was reduced to 1.0 mg dm^{-3} which is the detection limit of the indicator sensor for the benzene. The introduction of a chemical reagent extended this function. The retention time of benzene was 20 minutes after impregnation with TX-100. The absence of a second breakthrough in these two experiments indicates that chemisorption does not occur. In the case when the ACC was impregnated with copper tartarate ·two interesting points were observed. First, the protection time-adsorptive capacity of the cloth was even lower than in the case without impregnation. This result is in good agreement with our results in a static system (11). This suggests that the impregnation reagent blocked or occupied a part of the active area which was then nor available for the adsorption of benzene. However, the inlet concentration of benzene was reduced on passing by the cloth which can be ascribed to the process of chemisorption as indicated by the appearance of a second breakthrough point. Similar results were obtained with silver citrate and tartarate as impregnates and they where not used for further investigations.

In Figure 4 real recordings of data for toluene (C=110.9 mg dm^{-3}) passing through three characteristic ACCs (type 2) are presented: a) the ACC without impregnation, b) the ACC impregnated with TX-100 and c) the ACC impregnated with PVP. In these experiments PQCs were coated with 10 µg of TX-100, the flow-rate was 50 cm^3 min^{-1} and the temperature was 20° C.

It was observed that the ACC without any additional impregnation retained toluene 12 minutes. The introduction of TX-100 and PVP, as chemical reagents, extended this function to 20 and 30 minutes, respectively. In these experiments the frequency responses of both the indicator (A) and the referent (B) sensors are presented. During protection time the indicator sensor behaves as it feels nothing. This means that the outlet concentration of toluene is less than the detection limit of the indicator sensor for the toluene, that is 0.01 mg dm^{-3}. It can be calculated that 170 mg of

144

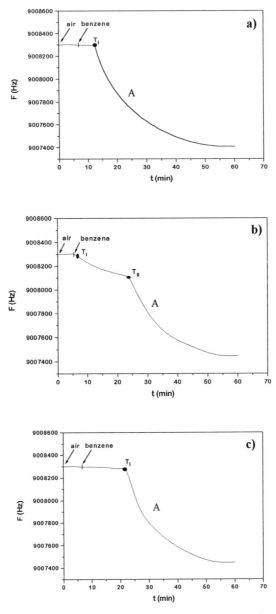

Figure 3. Breakthrough curve for benzene, C=319.0 mg dm^{-3}, Q=75 cm^{3}
min^{-1}, t=25°C. PQCs were coated with 10 µg of 4-AAP.

A-indicator sensor

T_I-first breakthrough point or penetration point, T_{II}-second breakthrough
point

a) ACC without impregnation; b) ACC impregnated with TX-100; C) ACC
impregnated with copper tartarate.

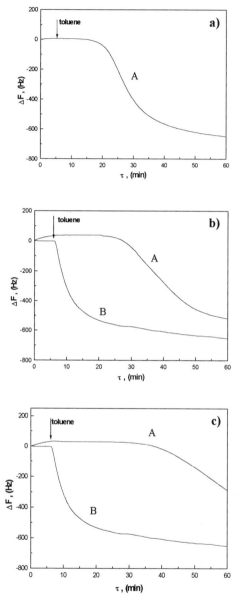

Figure 4. Breakthrough curve for toluene, C=110.9 mg dm^{-3}, Q=50 cm^3 min^{-1}, t=25°C. PQCs were coated with 10 μg of TX-100.
A-indicator sensor, B-referent sensor
a) ACC without impregnation; b) ACC impregnated with TX-100; C) ACC impregnated with PVP.

toluene is adsorbed onto 16 cm^2 of cloth during 30 min. The absence of a second breakthrough in these experiments indicates that chemisorption does not occur.

In Figure 5 real recordings of data for phenols (C=25.0 μg dm^{-3}) passing through three characteristic ACCs are presented: a) the ACC without impregnation, b) the ACC impregnated with 4-AAP and c) the ACC impregnated with PVP. In these experiments the PQCs were coated with 10 μg of 4-AAP, the flow-rate was 17.5 cm^3 min^{-1} and the temperature was 20° C.

Analyzing the response curves of the PQC sensor pair in Figure 5, the main advantages of this simple technique will be presented.

It can be estimated very quickly which type of basic ACC material is the best. In Figure 5a) curve B presents the frequency response of the referent sensor which corresponds with the inlet concentration of phenol. Curves A1$'$ and A2$'$ correspond to the two types of ACC. It can be concluded, without any additional knowledge about the detailed feature of these materials, that type 2 has better separation characteristics. The breakpoint depends upon the geometry, density and homogeneity of the activated carbon fibers. In the case of the ACC (type 2) the fibers are more compact. Curve A2$''$ corresponds with two layers of ACC-type 2.

The results presented in Figure 5b) were useful for an estimation of the effects of impregnation. The ACC impregnated with 4-AAP prolonged a little the penetration point (T_I) but not according to expectations. After the first breakthrough point, the phenols were bonded to the cloth by means 4-AAP. After depletion its activity the concentrations of the phenols in the main stream rose rapidly. This is indicated by the second breakthrough point (T_{II}). 4-AAP was chosen as a specific reagent for the spectrophotometric determination of phenols in the water. This reagent influenced the total reduction of phenol concentration.

In the experiments presented in Figure 5c) an ACC (type 1) was impregnated with two various technologies. One was washed after impregnation (A1), the other was used without washing in distilled water (A2). It is easy estimated that the ACC impregnated without washing (curve A2) is better for removal of phenols than the ACC impregnated without washing (curve A1). This method is convenient for the estimation of the impregnation technology.

From experiments presented in Figures 3-5 some very important estimations about ACC can be made: the quality of the basic material, the quality of the chemical impregnation agent, the mass of the chemical agent convenient for the chemisorption process and the penetration point or protection time. These demands are interesting for potential application, but from a fundamental point of view the most important is the possibility to estimate which chemical agents react during the sorption process and in that way contribute by chemisorption to the total sorption process.

Conclusion

A PQC sensor system can be successfully applied for the evaluation and characterization of various sorption materials in a continuous gas flow system. In this work the creation of a new sensor system for the characterization and evaluation of a sorption material in a flow system, based on coated piezoelectric sensors, was

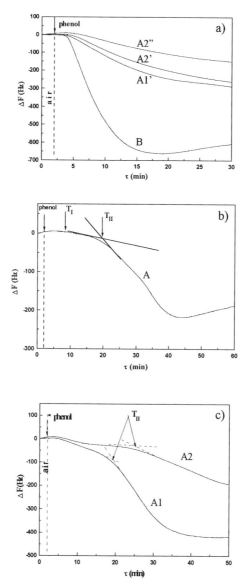

Figure 5. Breakthrough curve for phenols, C=25 μg dm⁻³, Q=17.5 cm³ min⁻¹, t=25°C. PQCs were coated with 1.0 μg of 4-AAP.

A-indicator sensor, B-referent sensor

T_I-first breakpoint or penetration point, T_{II}-second breakpoint

a) ACC without impregnation: A1'-one layer of ACC (type 1); A2'-one layers of ACC (type 2); A2"-two layers ACC (type 2);

b) ACC impregnated with 4-AAP;

c) ACC (type 1) impregnated with PVP: A1-ACC impregnated with washing; A2-ACC impregnated without washing.

148

attempted. For the detection of benzene 4-AAP exhibited the best feature as a coating and for the detection of toluene and phenols Triton X-100. The coated sensors exhibited reversible, stable, reproducible and reliable responses with high sensitivity and relatively long life time.

This simple methodology using a pair of PQC sensors enables the determination of the main criteria for the characterization of sorption materials: the time of protection, the penetration (breakthrough) point and the total capacity. It is concluded that the breakpoints and the breakthrough curves depend upon the nature of the pollutant and the sorbent material, particularly upon the chemical agent incorporated inside a material structure by impregnation. It has a strong influence on the retention time. Experiments with benzene, toluene and phenols, as representative organic air-pollutants, indicated that ACCs, impregnated with specific chemical agent for each pollutant, i.e. for benzene TX-100, and for toluene and phenols poly(vinyl pyrrolidone), possess the best features.

Acknowledgment

Author gratefully acknowledges the support given by the Serbian Research Council

References

1) Sauerbrey, G. Z.Phys. **1959**, 155, 206
2) King Jr., W.H. Anal Chem. 1964, 36, 1735
3) McCallum, J.J. Analyst **1989**, 114, 1173
4) Guilbault, G.G. Anal.Chim.Acta **1967**, 39, 260
5) Rajaković, Lj.V. J.Serb.Chem.Soc. **1991**, 56,521
6) Milanko, O., Milinković, S., Rajaković, Lj.V. Anal. Chim. Acta **1992**, 264, 43;
7) Milanko, O., Milinković, S., Rajaković, Lj.V. Anal. Chim. Acta **1992**, 269, 289
8) Thompson, M.; Kipling, A.; Dunkan-Hewitt, W.C., Rajaković, Lj.V. Analyst **1991**, 116, 881
9) Rajaković, Lj.V.; Ilić, M.R.; Jovanić, P.B.; Radošević, P.B. Carbon **1995**, 33, 10, 1433
10) Ilić, M.R.; Jovanić, P.B.; Radošević, P.B.; Rajaković, Lj.V. Separation Sci.Technol. 1995, 30, 2707
11) Ilić, M.R.; Jovanić, P.B.; Radošević, P.B.; Rajaković, Lj.V. J.Serb.Chem.Soc. **1995**, 60,507

Chapter 13

Electrochemical Oxygen Sensors: Principles and Applications

Naim Akmal[1] and Jay Lauer

Teledyne Eletronic Technologies-Analytical Instruments, 16830 Chestnut Street, City of Industry, CA 91749

This paper describes the construction, principle of operation and performance characteristics of galvanic electrochemical sensors for gas phase oxygen analysis. Purity of the electrolyte is very important to obtain accurate readings from the sensor at low levels of oxygen. Thickness of the gas permeable membrane plays a major role in the output and lifetime of the sensor. Temperature coefficient of the sensor has been found to be 2.5% per degree C. Selection of the appropriate noble metals depend on the level of oxygen and the background gases. Effects of anesthetic agents on the performance of the sensors has also been discussed. A new approach towards the development of a fast response time sensor (t_{90}<200 msec), that is comparable to the paramagnetic based technique one, has been described. Advantages of using rhodium metal cathodes in carbon dioxide background for automotive exhaust monitoring has been discussed briefly.

Oxygen gas is a very important target for chemical sensing because its presence controls various processes in the environment, in vivo, and in numerous industrial processes. Oxygen has a unique role in human existence. The nature and measurement of oxygen has, therefore, always been an important challenge to scientists. Modern technological advances both in industry and medicine have resulted in an increased demand for accuracy in oxygen measurement. The application of oxygen analyzers ranges from the measurement of the atmosphere inside an incubator containing a premature baby, to the

---

[1]Current address: Union Carbide Technical Center, 3200 Kanawha Turnpike, South Charleston, WV 25303.

©1998 American Chemical Society

monitoring of the oxygen content of process and combustion streams in giant chemical plants and refineries to ensure safe and economical operation. Monitoring of oxygen content in process gases in the semi-conductor manufacturing plant is essential to ensure high yield of the delicate memory chips. Most of the operations require oxygen levels to be below 5 ppb_v and in some cases at ppt_v levels. Use of a particular technique depends on the range of the oxygen to be monitored and requirement of the speed to be reported. Most processes require the response time to be under a minute at low ppb_v levels.

Existing oxygen sensors can be broadly grouped into two categories: (a) electrochemical and (b) optical. The electrochemical type include classical Clark type amperometric/polarographic devices (1) galvanic sensors (2) and solid-state electrolytic (e.g., Calcia/yttrium-doped zirconia) potentiometric sensors (3). This paper describes the principles of operation, various components, applications and recent developments in the area of galvanic sensors.

Principle of Operation

The electrochemical sensor is also called a Micro-Fuel Cell. A MFC is an electrochemical transducer, that converts one form of energy to another. In an electrochemical sensor, part of the energy is stored within and the other part comes from outside in the form of oxygen. The galvanic electrochemical sensor has three major parts: A cathode, an anode, and a pool of electrolyte in between. The reduction of oxygen molecules take place on the cathode and is known as the sensing electrode. The reaction at the cathode surface can be represented as:

$$4e^- + O_2 + 2H_2O \longrightarrow 4OH^- \tag{1}$$

In the above reaction, four electrons combine with one oxygen molecule to produce four hydroxyl ions. Presence of water is required for this reaction. This cathodic half-reaction occurs simultaneously with the following anodic half-reaction:

$$Pb + 2OH^- \longrightarrow PbO + H_2O + 2e^- \tag{2}$$

The anodic half-reaction presented above involves the oxidation of metallic lead (or other metals) into oxide and a transfer of two electrons for each atom of lead.

The net cell reaction thus being as follows:

$$2\,Pb\,(solid) + O_2\,(gas) \longrightarrow 2\,PbO\,(solid) \tag{3}$$

The current flowing in this cell can be used to monitor oxygen, provided the supply of the gas to the cathode (sensing electrode) was diffusion limited. For this purpose, a membrane of suitable porosity is interposed between the gas phase and the electrolyte in order to prevent saturation of the gas in the electrolyte.

Physical Construction

Galvanic type electrochemical sensors consume anode material during their operation. The sensor can be constructed in two ways. In the first method the electrolyte and the consumable anode can be recharged and in the second method the unit is designed free of any maintenance. Once all the anode is used up the sensor is disposed off requiring no maintenance during its lifetime. A typical disposable galvanic electrochemical sensor is shown in Figure 1. This sensor does not require replacement of the gas permeable membrane, electrolyte or anode. Also, the cleaning or polishing of the cathode on which the reduction of oxygen takes place is not required.

Electrolyte. The most commonly used electrolyte in electrochemical sensors is a 10-15% aqueous solution of potassium hydroxide. With appropriate precautions, electrochemical experiments are possible in almost any medium, and, indeed, experiments have been described in molten naphthalene, glasses, and *in vivo* in rats' brains! Table I provides a list of some widely used media for electrochemical sensors (4).

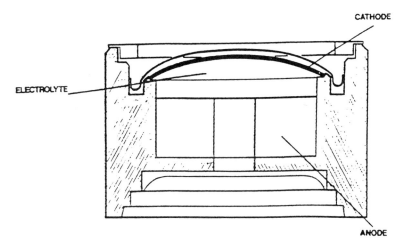

Figure 1. Schematic representation of a disposable galvanic oxygen sensor.

In order to have a clean reaction for oxygen and its detection at ppb_v levels, the electrolyte should be free of any impurities including any organic species. Therefore, salts should be of the highest available purity and solvents should be carefully purified. Water should be distilled thrice to remove any traces of impurities and sometimes pre-electrolyzed for use in solutions for sensors used for low (ppb_v) levels of oxygen detection.

Anode. The most commonly used material for anode is pure metallic lead. Lead metal in granular form is pressed into pellets inside the sensor body in the presence of caustic solution. Cadmium metal paste can also be used as an alternative to lead. Cadmium has an advantage, since it has fewer oxidation states than lead. The oxidation products (using cadmium as anode) are much cleaner and do not interfere with the reduction step of oxygen. However, a major disadvantage using cadmium is its toxicity and the electrode preparation requirement before use. Also the metallic cadmium turns into cadmium oxide during its exposure to oxygen. Therefore, additional step is required in order to reduce the oxidized cadmium back to cadmium metal. This step is performed by applying the necessary potential with respect to a platinum electrode inside the electrolyte solution. After the reduction step, the anode is kept inside the electrolyte to prevent its premature oxidation.

Cathode. The reduction of oxygen takes place on the cathode. In order to avoid any kind of interference during the oxygen reduction step, only noble metals such as gold, silver, platinum, etc. are used for this purpose. To keep the cost low, the disposable galvanic sensors have cathodes made up of perforated brass protected by a thick layer of noble metal. During the lifetime of the sensor, the products generated such as lead oxides have the potential to form crystals on the cathode. The deposition of the oxide layer reduces the output of the sensor. It has been found that the selection of the noble metal depends on the nature of the application. For low levels (ppm_v and ppb_v) of oxygen detection use of silver plated cathodes is recommended. Rhodium is preferred as a cathode metal in a carbon dioxide background, since it prevents the formation of lead carbonate on the cathode surface. This is highly desired in cases where a continuous measurement of oxygen is required in a carbon dioxide background.

Membrane. A polymer membrane is placed over the cathode that is permeable to gases but impermeable to the electrolytes. Hence, the P_{O2} gradient is mainly confined across the membrane. As a consequence, porosity and thickness of the membrane are very important design parameters for accurate sensor operation in any individual oxygen concentration range. Teflon is the most commonly used membrane for this application. The membrane selected must be free of any pin holes and wrinkles. The thickness of the membrane varies from 3 mil to 1/8 mil depending on the application. A sensor made out of thick membrane has a lower output and has a longer response time. Also, the sensor's lifetime depends on the thickness of the membrane. A thicker membrane cuts down the diffusion of oxygen to the cathodes. Polyethylene has also been used as a diffusion membrane. Due to the presence of pinholes and high temperature coefficient, polyethylene is not a good choice.

Oxygen Measurement.

In reaction (1) it can be seen that four electrons are transferred for each oxygen molecule undergoing reaction. The rate at which oxygen molecules reach the surface of the cathode determines the electrical output. This rate is directly proportional to the concentration of oxygen in the gaseous mixture surrounding the sensor. Figure 2 shows the concentration versus output for a galvanic sensor. The output is linear,

Table I. Common Solvents and Media for Electrochemical Experiments

Water
Aqueous solutions of many salts and/or complexing agents at various pH.
Buffered and unbuffered media.

Other protonic solvents
e.g. acetic acid, methanol, ethanol, glycerol, ethylene glycol.

Aprotic solvents
e.g. acetonitrile, dimethylformamide, dimethylsulphoxide, sulfur dioxide, ammonia, propylene carbonate, tetrahydrofuran etc.

Mixed solvents
Mixtures of water and ethanol , mixtures of water and glycerol, mixtures of water and ethylene glycol. These media may be buffered or unbuffered depending on the medium in which oxygen is being measured.

Molten salts
e.g. NaCl, KCl/LiCl eutectics etc.

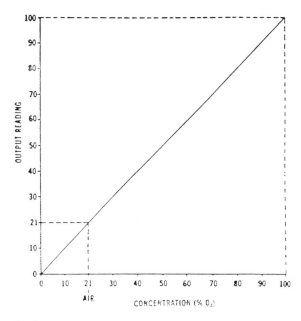

Figure 2. Output vs. concentration plot for a galvanic oxygen sensor.

exhibiting an absolute zero. This shows that in the absence of oxygen the sensor produces no output (5). Given these two conditions (linearity and absolute zero), single point calibration is possible. The oxygen concentration in air (20.945%) is a very convenient reference gas to use, and is accurate up to three or four significant figures throughout the entire earth's surface.

In calibrating a given oxygen analyzer, it is always best to use a calibration gas whose oxygen content is near or equal to the full scale reading. In Figure 2, both air and 100% oxygen are shown as potential calibration gases. Theoretically, any known oxygen concentration can be used. Air and 100% oxygen were chosen because they are the easiest and most economical to obtain.

Galvanic technology (electrochemical) based oxygen sensor measures the partial pressure of oxygen in the sample gas mixture, although most of the sensors are set up to display a percent reading. Readouts in percent are permissible only when the total pressure of the gas being analyzed does not change.

To illustrate the foregoing, assume that a sensor with a percent readout was calibrated in air at sea level, the span control would be adjusted until the sensor reading was 20.94% (assuming that degree of readability was available). If the sensor was then elevated to higher and higher altitudes, the reading would begin to drop and continue to do so as long as the sensor was being raised to a higher altitude. The percent of oxygen in the atmosphere is the same regardless of the altitude, i.e., the ratio of oxygen molecules to the total number of gas molecules does not change, the molecules merely get farther apart from each other. The sensor's output is a function of the number of molecules of oxygen per unit volume, which is decreasing as the sensor is raised in altitude. Figure 3, illustrates how the output of the sensor increases when the total pressure is increased (the opposite of the previous illustration). In this case, as with the previous one, the percent oxygen does not change.

The output of a sensor changes with a change in the oxygen concentration, but can also change with a change in the temperature of the gas. The rate at which oxygen molecules hit the cathode is governed by diffusion through the sensing membrane and the film of the electrolyte between the membrane and the cathode. The output of the sensor varies at approximately 2.5% per degree C. Therefore, the sensor has a positive temperature coefficient of 2.5% per degree C. To compensate for this effect, negative temperature coefficient thermistors are used. Figure 4 shows the manner in which this temperature compensation is effected. In most instruments (using galvanic sesnors), the compensation curve is matched with an accuracy of ±5% or better. The resultant output function then, is independent of temperature and varies only with changes in the oxygen concentration. Figure 4 also shows two simple circuit diagrams where the sensor's output is temperature compensated, one involving no amplifier and the other showing a single stage of amplification.

Galvanic Sensors for Parts-per-Billion Oxygen Detection.

Galvanic sensor technology has successfully been used to detect ppb_v levels of oxygen in process and nitrogen gases for semi-conductor industry (6). In this application, the cathode was made out of a metal catalyzed gas diffusion electrode. The signal output from the gas diffusion electrode was approximately 600 times greater than the signal

155

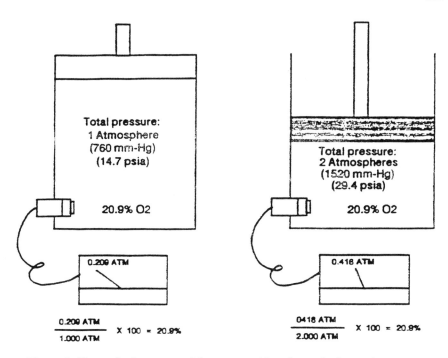

Figure 3. Change in the output of the sensor with a change in the total pressure.

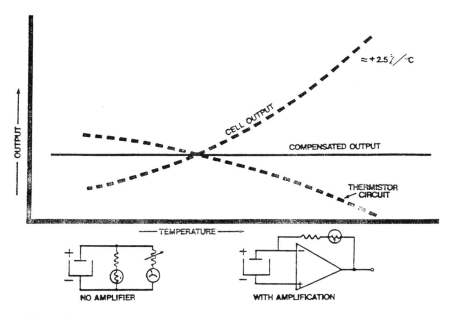

Figure 4. Temperature compensation scheme for a galvanic sensor. Circuit diagrams where sensor's output is temperature compensated.

generated from a smooth metal electrode. In order to see the signal generated for low ppb$_v$ levels of oxygen, the pool of electrolyte was continuously sparged with oxygen free nitrogen gas. Figure 5 shows the sensor designed for this application. The body of the sensor was made of clear acrylic and filled with a 10-15% aqueous potassium hydroxide solution. Figure 6 shows the plot of the calculated oxygen concentration versus the experimental values obtained. There is an excellent agreement between the calculated and experimental values; the experimental values are within ±1 ppb$_v$ of the calculated values.

Galvanic Sensors for Medical Applications.

It is necessary to monitor oxygen concentration in the patients breath during the introduction of anesthetic agents. The sensors should not interfere with the actual oxygen readings Table II shows the maximum interference observed by the sensors in the presence of various anesthetic agents.

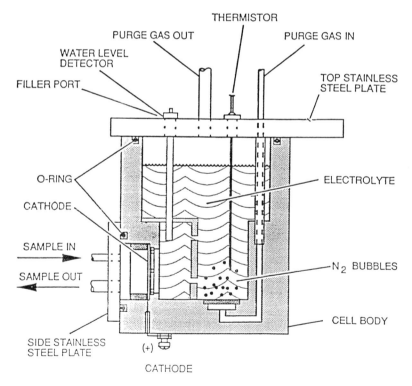

Figure 5. Schematic representation of a trace oxygen (ppb$_v$) sensor. (Reproduced with permission from ref. 6. Copyright 1991 Instrument Society of America).

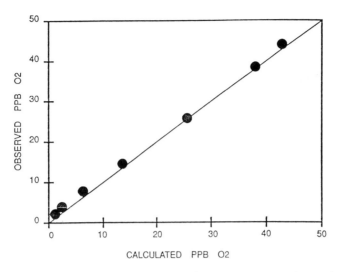

Figure 6. Plot of the theoretical vs. observed oxygen concentrations using a gas diffusion electrode sensor; theoretical (solid line) and reading from the sensor (filled circles). (Reproduced with permission from ref. 6. Copyright 1991 Instrument Society of America).

Table II. Effects of Anesthetic Agents on Oxygen Sensors

Agents	Concentrations (%)	Reading Error (%O_2)
Halothane	6	1.1
Isoflurane	6	0.9
Enflurane	6	0.5
Sevoflurane	6	1.1
Desflurane	20	0.5

Reading error in terms of oxygen concentration is under 1.25% in all cases. Thus galvanic sensors have a major application in the operation rooms, since they are not subject to any interference from anesthetic agents.

There is a great demand for a fast response oxygen sensor (<200 msec) in the medical industry to monitor the oxygen level in breath of patients. The currently available technology is based on the paramagnetic technique and is very expensive (>US$10,000). A new galvanic sensor was developed that has a response time of less than 200 msec at percent levels of oxygen. The sensor was made of a thin layer of gold (500°A) deposited on a 1/8 mil thick Teflon membrane. The response time for this sensor is shown in Figure 7. The electrolyte used was a mixture of aqueous and organic solvents and caustic.

Galvanic Sensors for Automotive Exhaust Monitoring.

The increasing number of automobiles are creating a major health problem to the residents of big cities by producing smog. Smog is caused by the exhaust gases coming out of the tail pipe of automobiles. The efficiency of a gasoline engine can be increased by managing the fuel to air ratio. The galvanic sensor is used to detect oxygen level coming out of an automobile's tail pipe. The exhaust gas contains a high concentration of carbon dioxide along with other gases. It has been found that the presence of carbon dioxide in the sample stream produces a layer of lead carbonate on the surface of the cathode leading to a drop in the ability of the cathode to reduce oxygen. It has been found that use of platinum or rhodium as noble metals improves the performance of the sensor in a carbon dioxide background. Cathodes plated with silver and gold metals stop working after few hours of usage, whereas cathodes plated with rhodium continued to work more than 100 h. Figure 8 shows this ability of the rhodium based cathode to reduce oxygen in carbon dioxide background. Rhodium also has been found to display up to 99.3% linearity when the sensor was exposed to a series of oxygen concentrations ranging from 0 to 100%. Cathodes plated with gold showed maximum non-linearity under similar conditions. Figure 9 shows the linearity data for various cathode materials.

Conclusion

Galvanic electrochemical sensors are easy to operate and are relatively inexpensive. These sensors have excellent sensitivity (1 ppb_v O_2) and accuracy, especially when the sensing electrode (cathode) is a metal catalyzed gas diffusion electrode. The fast responding sensors are finding additional applications in the medical and automotive industries due to the low cost of ownership. The lifetime of the sensors can be enhanced by modifying the electrolytes and also the diffusion pattern of the oxygen to the sensing membrane.

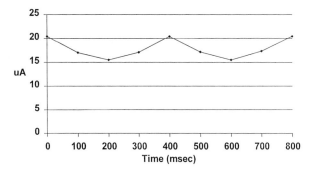

Figure 7. Response time of a fast oxygen sensor for percent oxygen concentrations.

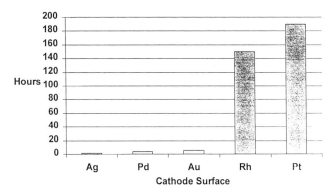

Figure 8. Continuous monitoring of oxygen in 16% CO_2/balance N_2 gas using galvanic sensors made of various noble metal cathodes.

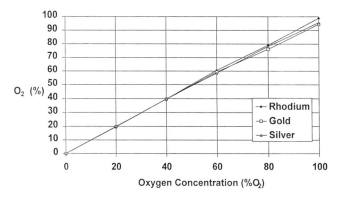

Figure 9. Linearity for various noble metal coated cathodes from 0 to 100% of oxygen.

160

References

1. Clark, L. C. Jr. *Trans. Am. Soc. Ant. Int.* Organs **1956**, *2*, 41.
2. Mancy, K. H.; Okun, D. A.; Reilley, C. N. *J. Electroanal. Chem. Interfacial Electrochem.* **1962**, *4*, 65.
3. Siebert, E.; Fouletir, J. *Ion-Selective Elect. Rev.* **1986**, *8*, 133.
4. Greef, R.; Peat, R.; Peter, L. M.; Pletcher, D.; Robinson, J. *Instrumental Methods in Electrochemistry*; Ellis Horwood Limited: West Sussex, UK, 1985, pp 364.
5. Yamazoe, N.; Miura, N.; Arai, H.; Seiyama, T. *Proceedings of TRANSDUCERS' 85*, International Conference on Solid State Sensors and Actuators: Philadelphia, PA, **1985**, pp 340.
6. Razaq, M. *ISA Proceedings'* 91: **1991**. pp 1739-1748.

Chapter 14

Selectivity of an Anatase TiO$_2$-Based Gas Sensor

Sheikh A. Akbar, Lora B. Younkman[1], and Prabir K. Dutta

Center for Industrial Sensors and Measurements, The Ohio State University, Columbus, OH 43210

The analysis of electrical data combined with X-ray and XPS observations support a sensing mechanism involving CO adsorption and ionization on the anatase titania surface. In the case of H$_2$, because of slow recovery, a diffusion-controlled mechanism is suggested. The dependence of the grain and grain-boundary resistances and capacitances to CO concentration remained unaffected by the Y$_2$O$_3$ addition, indicating that the second phase does not change the underlying sensing mechanism. The addition of Y$_2$O$_3$ is likely to block the diffusion of hydrogen into the bulk of TiO$_2$, thus suppressing the response toward H$_2$. The addition of Fe greatly enhanced the sensitivity toward CO. This indicates that oxygen vacancies created due to Fe^{3+} substitution on Ti^{4+} sites may play a major role in the enhanced sensitivity.

Since the advent of metal oxide gas sensors three decades ago, substantial research and development efforts have focused on their application for gas detection. In these sensors, specific gases are detected and quantified by changes in the electrical resistance of the sensing material due to the adsorption of reactive gases on the surface. Numerous metal oxide ceramics/semiconductors have been employed as sensing devices for a variety of gases, including CO, CO$_2$, H$_2$, H$_2$O, NH$_3$, SO$_x$ and NO$_x$, with varying degree of commercial success *(1-3)*.

Ceramic gas sensors are typically fabricated as sintered porous pellets or thick films, in which the resistance of the material depends very strongly on gas adsorption *(4,5)*. In n-type semiconductors such as SnO$_2$, a two-step process has been suggested for the electrical response to reactive gases *(6)*. First, oxygen from the ambient adsorbs on the exposed surface of the grains, and extracting electrons from the material, ionizes to O$^-$ or O^{2-}, with O$^-$ being dominant at higher temperatures *(7)*. The extraction of electrons leads to the formation of a depletion region at the intergranular contact. Thus, to begin with, the sensor surface is highly resistive. A reducing gas such as CO then reacts with the adsorbed oxygen, and injects electrons

---

[1]Current address: Delphi Energy and Engine Management Systems, 1300 N. Dort, Highway, M/C 485-220-130, Flint, MI 48556.

©1998 American Chemical Society

back into the conduction band, thereby decreasing the barrier height and resulting in the increase in conductivity. For most materials, the barrier height is not identical at each intergranular contact, but some distribution exists *(8)*. In such a case, the charge carrier flows through a connected path of lowest barrier heights, which controls the sensitivity. By adding a uniformly distributed second phase to the material, this path is interrupted, and the conductivity and resulting sensitivity can be altered.

While ceramic sensors offer advantages of small size, low cost and *in-situ* monitoring capabilities for high temperatures, they typically suffer from nonselectivity. They respond to a wide spectrum of reducing gases, therefore making it difficult to assign the signal to a particular gas of interest in a mixture of other interfering gases. Selectivity is generally achieved through the inclusion of metal or metal oxides to the sensor material. Though the addition of a second phase seems to be a common approach for achieving selectivity, the selection of the additive becomes a challenge because of the lack of a basic understanding.

Recent work in the authors' laboratory has led to the development of an anatase titania (TiO_2)-based ceramic material as a reliable and rugged CO and/or H_2 gas sensor at high temperatures*(9-11)*. This article reports the results of some preliminary studies on the role of the second phase toward the selectivity of the anatase-based sensor.

Experimental Procedure

All samples in the present investigation were fabricated from high purity TiO_2 (X-ray pure anatase phase, J.T. Baker Inc., Phillipsburg, New Jersey), Y_2O_3 (Alfa Ventron, Alfa Products, Danvers, MA) and fine-grain Fe (Baker-Analyzed reagent grade) powders. The mixed powder was ball milled with ZrO_2 media in isopropanol for 24 h to achieve a uniform mixture. The slurry was subsequently dried at 200 °C for 8 h, homogenized in an agate mortar and sieved through a 325-mesh screen.

The resulting powder was mixed with 1-heptanol, printed through a 325-mesh screen onto an alumina substrate and dried at 200 °C for 2 h, leaving a porous thick film. Gold wires (0.25 mm diameter, 100 mm long) were attached to the film with conductive gold paint (Engelhard H1109, New Jersey) to form electrical leads. The films were then heat treated at 850 °C for 4 h. Since the anatase to rutile phase transformation has been observed to occur at 900 °C in ambient air, this heat treating schedule was selected to avoid any rutile formation. Phase identification by X-ray diffraction (XRD) analysis verified that the anatase phase was retained.

Both the two-probe and four-probe measurement techniques were used to obtain the electrical resistance. There was no observable difference in the measured resistance between the two techniques. The data reported in this paper are based on the two-probe method. Each test sample was positioned in the uniform temperature zone of a precalibrated horizontal Lindberg furnace, with temperature fluctuations of less than 1 °C. Testing temperatures ranged from 500 to 800 °C in environments of various levels of CO in background N_2, accurately maintained with a Sierra 5-channel control unit.

The dc resistance was measured using a Keithley multimeter (model 169) and was simultaneously monitored on strip chart recorder. Only after the resistance had reached a steady value identified by a plateau in the chart recorder plot, the CO level was changed. At each CO level, it took 3-10 minutes to reach the steady value. The large volume of the furnace tube (~20 L) employed in this study is responsible for the slow response. Such a large volume makes the mixing and equilibration of CO with background N_2 a time dependent process. Thus, there is a definite time lag between the entry of the gases into the furnace and mixing, adsorbing on the sensor surface, and generating the electrical signal.

The ac electrical data were acquired over the frequency range of 5 Hz to 13 MHz with an impedance analyzer (Model 4192A Hewlett-Packard, Tokyo, Japan). Necessary electrical parameters were extracted through a complex nonlinear least-square (CNLS) curve-fitting software program developed in the authors' laboratory. Unlike most commercial software, this program is capable of extracting all necessary parameters without assuming or simulating any equivalent circuit configuration a *priori.(12)* The acquired ac electrical data were plotted in the four complex planes: impedance (Z^*), admittance (Y^*), modulus (M^*) and capacitance (C^*). In the present case, semicircular relaxations were observed only in the impedance plane and not in the parallel representations (C^* and Y^*). Grain and grain-boundary (intergranular) resistances and capacitances were obtained from the impedance-plane plot.

Results and Discussion

Figure 1 shows the dc resistance of a TiO_2 thick film sample on CO concentration, at 500, 600 and 700 °C. The sensor showed good reversibility with respect to increasing and decreasing levels. The sensor could also be completely regenerated, merely by shutting off the CO gas, without exposing it to any oxidizing atmosphere such as air or oxygen. This is in marked contrast to several other gas sensors based on semiconducting oxides, where regeneration of the sensor requires the oxidation by either air or oxygen *(13)*.

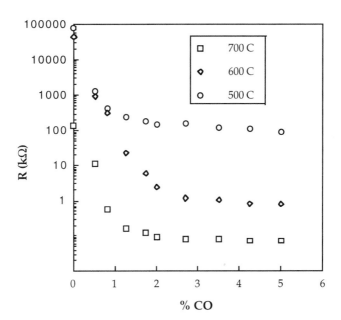

Fig.1   Resistance (dc) isotherms of TiO_2 thick films as a function of CO gas concentration.

The ac electrical data were acquired at 500 °C for various CO concentrations. Figure 2 shows impedance plane plots in (a) the absence of CO and (b) 1000 ppm CO in background N_2. The two distinct relaxations represented by the low and high frequency semicircular arcs can be attributed to the lumped grain-boundaries and lumped grains, respectively. The grain (R_1) and grain-boundary (R_2) resistance values were extracted from the Z*-plane plots for the various CO concentrations. Both R_1 and R_2 decreased with increasing CO concentrations with the latter showing a rapid rate of reduction. The grain-boundary capacitance (C_2) was observed to increase rapidly upon the introduction of CO, reaching a saturation value at higher concentrations. Such an increase in the capacitance was attributed to a decrease in the width of the depletion region at the intergranular contact (13).

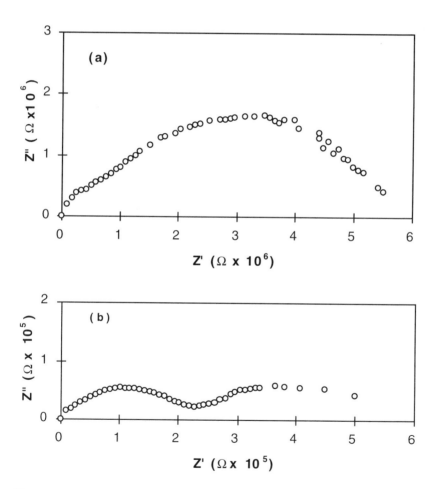

Fig.2 Impedance plane plots for TiO_2 thick film at (a) 0 ppm and (b) 1000 ppm CO in N_2, measured at 500 °C.

Based on these observations, a simple sensing mechanism has been proposed *(13,14)*. The adsorbed CO ionizes on the surface, releasing an electron into the conduction band. Upon increasing CO concentrations, more charge carriers are generated, and resistance decreases. Such an increase in conduction band electrons effectively lowers the depletion region thickness as well as the intergranular barrier height. This decrease in electrical thickness results in an increase in the grain-boundary capacitance. Thus, the observed capacitance behavior provides support for the proposed depletion layer controlled sensing mechanism.

X-ray diffraction was employed for phase identification of the titania surface, both before and after exposure to CO gas. In both cases, the X-ray spectra revealed anatase TiO_2 to be the only phase present. No trace of Ti_2O_3, TiO or any other suboxide of titanium was revealed in the diffraction pattern. In addition, no color change of the film surface to the black TiO phase or purple Ti_2O_3 phase was observed. The films remained white in all cases after exposure to CO gas. XPS analysis performed both before and after exposure to CO also confirmed that titania surface did not reduce to a lower oxide. Analysis based on the ac electrical data combined with X-ray and XPS observations tend to support a sensing mechanism involving CO adsorption and ionization on the titania surface, and not an oxidation-reduction type reaction as observed in most oxide-based sensors *(13)*.

The anatase TiO_2 was also found to be a good H_2 sensor as illustrated in Figure 3. The reversibility characteristics in the case of H_2, however, was not as good as in the case of CO; it took more time for the resistance to rise upon decrease in H_2 gas concentration. Regeneration of the sensor was very sluggish after shutting off the H_2

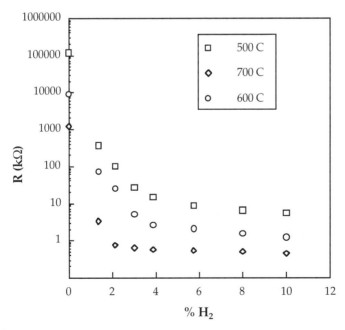

Fig.3 Resistance (dc) isotherms of TiO_2 thick films as a function of H_2 gas concentration.

gas. Based on these characteristics, a diffusion controlled mechanism was proposed for H_2 sensing *(16)*. This mechanism involves the dissociation of H_2 into H atoms on the TiO_2 surface. These H atoms then diffuse into the bulk of TiO_2 and ionize to produce conduction electron and interstitial protons. When H_2 gas concentration is decreased, H atoms have to diffuse back to the surface leading to a poor reversibility and slow recovery.

Figure 4 shows the response of a thick film sensor made form TiO_2-10 wt% Y_2O_3 (TY) to CO and H_2. The sensitivity is defined as the relative dc resistance, R/R_o, R_o being the resistance in background nitrogen. The results reveal that the sensitivity of the two phase mixture for hydrogen is reduced, while that for CO remains unaffected. It is believed that Y_2O_3 forms a diffusion barrier for H atoms, thus suppressing the sensitivity toward H_2 gas.

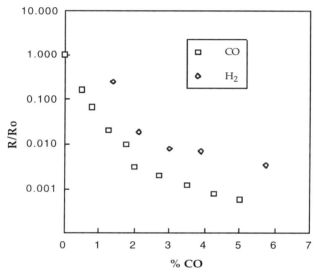

Fig. 4 Sensitivity of TY (TiO_2-10 wt% Y_2O_3) to CO and H_2 gases; R_o is the dc resistance in background nitrogen.

In order to explore the role of the second phase, the ac electrical data were acquired on a TY thick film sample. The data were analyzed using the complex plane plots as described earlier. With the increase of CO concentration, the grain-boundary resistance and capacitance for TY also showed similar dependence to that of the single-phase TiO_2. These results suggest that the addition of Y_2O_3 does not appear to change the proposed depletion layer controlled sensing mechanism.

Although the addition of Y_2O_3 does not appear to change the underlying sensing mechanism, the addition of Fe in TY was found to give dramatic effect. The addition of Fe greatly enhanced the sensitivity toward low concentration of CO as depicted in Figure 5. Fe^{3+} substitution on Ti^{4+} sites is expected to create oxygen vacancies for electrical compensation *(15)*. It is believed that these oxygen vacancies may play a major role for the enhanced sensitivity toward CO; the exact role of theses vacancies is not clear at this stage. The other possibility is that Fe_2O_3 may play a role as a catalyst. Further studies toward the understanding of the role of Y_2O_3 and Fe_2O_3 should lead to the optimization of sensitivity and selectivity of theses sensors.

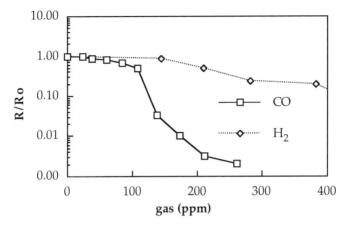

Fig.5   Response of a TY (TiO$_2$-10 wt% Y$_2$O$_3$) thick film impregnated with Fe to low levels of CO and H$_2$ at 500 °C.

Acknowledgments

This work was supported by the National Science Foundation grants (DMR-9503429 and EEC-9523358) with matching support from the Edison Materials Technology Center (EMTEC) of the state of Ohio and the Edward Orton Jr., Ceramic Foundation (Westerville, OH). Technical discussions with Dr. C.C. Wang and his help with the preparation of the manuscript is gratefully acknowledged.

Literature Cited

1. Göpel, W; Schierbaum, K. *Sensors: A Comprehensive Survey*, 1991, Vol.2, 430.
2. Azad, A. M.; Akbar, S. A.; Mhaisalkar, S. G.; Birkefeld, L. D.; Goto, K. S. *J. Electrochem. Soc.*, 1992, 139, 3690.
3. Watson, J.; Ihokura, K.; Coles, G. S. V. *Meas. Sci. Tech.*, 1993, 4, 711.
4. Morrison, S. R. *Sensors and Actuators,* 1987, 11, 283.
5. Moseley, P. T. *Sensors and Actuators B,* 1992, 6, 149.
6. Windischmann, H.; Mark, P. *J. Electrochem. Soc.*, 1979, 126, 627.
7. Mark, P. *J. Chem. Phys. Solids,* 1968, 29, 689.
8. Sukharev, V. Y. *J. Chem. Soc. Faraday Trans.*, 1993, 89, 559.
9. Birkefeld, L. Azad, A. M.; Akbar, S. A. *J. Am. Ceram. Soc.*,1992, 75, 2964.
10. Azad, A. M.; Akbar, S. A.; Younkman, L. B. *Interface*, 1994, December, 31.
11. Akbar, S. A.; Azad, A. M.; Younkman, L. B. *A Solid-State Gas Sensors for Carbon Monoxide and Hydrogen*, patent # 5,439,580, August (1995).
12. Zhu, W.; Wang, C. C.; Akbar, S. A.; Asiaie, R.; Dutta, P. K.; Alim, M. A. Jpn. J. Appl. Phys., 1996, 35, 6145.
13. Akbar, S. A.; Younkman, L. B. *J. Electrochem. Soc.*, 1996, in print.
14. Birkefeld, L. D.; Azad, A. M.; Akbar, S. A. *J. Am. Ceram. Soc., 1992, 75,* 2964.
15. Hishita, S.; Mutoh, I.; Koumoto, K.; Yanagida, H. *Ceramics Int.*, 1983, 9, 61.

Chapter 15

The Potential of Piezoelectric Sensors for Characterization of Activated Carbon Cloth Applied in Adsorption of Phenols from Air

Lj. V. Rajakovic and A. E. Onjia

Department of Analytical Chemistry, Faculty of Technology and Metallurgy, University of Belgrade, Karnegijeva 4, POB 494, 11 000 Belgrade, Yugoslavia

Characterisation of activated carbon cloth (ACC) applied for phenols adsorption from the air was performed by a pair of coated piezoelectric quartz crystals as sensors. ACC cloth was impregnated with specific organic compounds, and its sorption properties were monitored in a dynamic flow system. The concentration of phenol was permanently controlled using one coated crystal as the referent inlet and the other one as the indicator outlet sensor. Frequency responses of both sensors in the system provided the data for considering the breakthrough points, and the other sorption parameters. It was concluded that ACC impregnated with specific compounds specially with poly(vinyl pirrolidone) have an optimal capacity from the point of view of both physical adsorption and chemisorption. Also, piezoelectric quartz crystals coated with Triton X-100 and 4-amino-antipyrine were the sensitive tools for the monitoring of sorption process of phenols from the air onto the impregnated ACC.

Coated piezoelectric quartz crystal (PQC) has been successfully used as a gas-phase microgravimetric and chemical sensor for the assay of various environmental organic traces in the air (1-5). High sensitivity, raggedness and low cost of this device makes it convenient for use in analytical chemistry and environmental study. A bulk acoustic wave sensor in the thickness shear mode (BAW/TSM) has been used extensively in the phase and much less in the liquid phase (6). King (2) introduced the piezoelectric chemical sensor and addressed almost every aspect of its use. Most of the work done subsequently was aimed at finding the appropriate coatings for the assay of the most common air pollutants. In this work a new application of PQC was investigated and applied for the characterisation of various sorption materials, specially ACC.

Various types of ACC with and without impregnation were used as active and selective sorpion materials (7-9). ACC was activated by introducing some

©1998 American Chemical Society

specific chemical agents. In this way, it is possible to separate toxic air pollutants efficiently by their bonding to an or more active chemical agent inside the cloth. Impregnation usually has been done by using some specific compound such as Triton X-100, (TX-100), poly(vinyl pyrrolidone), (PVP); 4-amino-antipyrine, (4-AAP), formaline and copper tartarate.

A special task in this work was to remove phenols by the ACC and to detect concentration of phenols during separation process by the PQC. This task was solve by the application of a pair of PQCs. One of PQCs was used as the referent inlet sensor and the other was used as the indicator outlet sensor. In that way the inlet concentration of the air-pollutant was permanently controlled, the decrease in the pollutant concentration due to sorption onto the ACC was monitored and the penetration point was determined.

Experimental Section

The basic materials in this research were microporous the ACC with a specific surface of 1500 m^2 g^{-1}, made by carbonisation and activation of a standard production type viscous rayon, produced by AJMV, Kijev. Tartarates and citrates of copper and silver as impregnation reagents were used. Concentration of copper-tartarate solution was 0.1 mol dm^{-3}, and they were made from Merck. For the specific reagents solution as TX-100, PVP, 4-AAP and formaline concentration of solutions were 0.1%. The experimental procedure for impregnation and characterisation of carbon cloth by classical methods have been described elsewhere (7-9).

An apparatus used for generating test vapours, exposing the ACC (16 cm^2) as sorption material and the coated PQCs as sensors as well as characteristic points on idealised response curves were evaluated and described in our first paper presented in the ACS Symposium (10). The coated 9 MHz, Al-plated PQCs were exposed in sequence to the pure air, the air with the phenols sample and the pure air again. Concentration of·phenols in the air was constant 25.0 µg dm^{-3}. The concentration of phenol was tested by GC (5). An optimal flow rate of 17.5 cm^3 min^{-1} with respect to sensor response and recovery time was used. All the experiments were performed at room temperature.

Results and Discussion

Many factors govern the behaviour of an oscillating PQC. Mass transport of a gaseous sample through the coating is affected by surface roughness, thickness, homogeneity and morphology. However, the sensitivity of a sensor depends primarily on the mass and type of the coating (6). The data presenting in Figure 1a for the interaction of phenol and three coatings indicate that each type of coating possesses its own particular mass sensitivity. The choice of various compounds as coating was based on empirical and theoretical knowledge. The choice of 4-AAP is based on the fact that this specific reagent is usually used for the classical spectrophotometric determination of phenols in a liquid medium (5). The use of polymeric compounds such as TX-100 and PVP is based on the possibility of its binding to the Al-plated crystal, giving stability to the sensor, and its reactivity towards various gaseous compounds (4). It was concluded that PQC coated with TX-100 is the most sensitive tools for the monitoring of phenol concentration in the air.

The sensitivity of the sensor was affected by the amount of coating deposited on the crystal. An optimal mass of coating was 10.0 μg that is a compromise between the maximal sensitivity and the rate of response and recovery of the sensor. The detection limit of this sensor for the phenols is 0.05 μg dm^{-3}.

Next step was to choose the best chemical agent for the impregnation of the ACC which could bond phenols chemically and in that way improve the effect of removal process. In Figure 2 real recordings of data for the phenols (C=25.0 μg dm^{-3}) passing through four characteristic ACCs are presented: a) the ACC impregnated with copper-tartarate, b) the ACC impregnated with formaline c) the ACC impregnated with TX-100 and d) the ACC impregnated with 4-AAP. In these experiments PQCs were coated by 10.0 μg of TX-100.

These results confirmed that activated carbon cloth can be characterised by this simple procedure based on a pair of piezoelectric sensors. The time of protection until penetration point (T$_I$) is the main criteria for this type of materials and the answer can be obtain with this simple methodology. Also, the time when a chemical reagent ends its activity can be estimate by the second breakthrough time T$_{II}$.
It was observed that the ACC without impregnation were not active, phenols was retained-adsorbed only a few minutes. By introducing a chemical reagent various effects were observed. Chemical agents as copper tartarate (Fig.2a) formaline (Fig.2b), TX-100 (Fig.2c) and 4-AAP (Fig.2d) were not effective. Some of them as TX-100 and 4-AAP had no influence to the penetration point, but they increased the total capacity of the cloth. Appearance of the second breakthrough point in these experiments indicates that chemisorption takes places.

In Figure 3 a real recording of data for the phenols (C=25.0 μg dm^{-3}) passing through the ACC impregnated with PVP is presented. In this experiment PQCs were coated by 10.0 μg of 4-AAP.

Curve B in Figure 3 presents frequency response of the referent sensor which corresponds with the inlet concentration of phenols. Curve A corresponds with the outlet concentration of phenols.
The experiments presented in Figures 2 and 3 were useful for the estimation of the effects of impregnation. It is obvious that the best material for the phenol removal from the air is the ACC impregnated with PVP. It possesses the highest time of protection - more than 60 min. It should be note that the initial concentration was reduced about 90% from the very beginning. Phenols are widespread toxic organic pollutants of the atmosphere. Their content in industrial areas must be monitored and strictly controlled. The application of a pair of PQC sensors for such compounds detection and removal is of environmental interest.

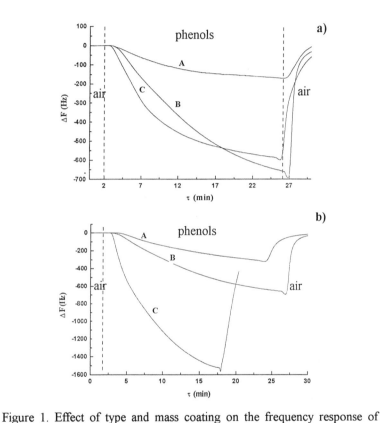

Figure 1. Effect of type and mass coating on the frequency response of 9MHz Al-plated PQC exposed to phenols.
a) PQCs were coated with 10 µg of the coatings; Coatings: A-4AAP; B-TX-100; C- PVP; b) PQCs were coated with TX-100, A- 5 µg, B-10µg, C- 40µg.

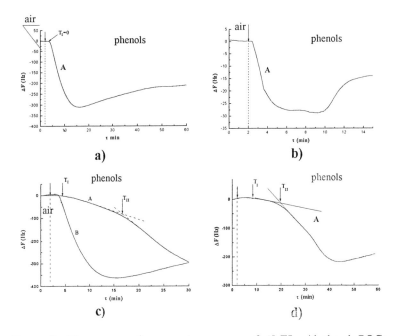

Figure 2. Time-course frequency responses of 9MHz Al-plated PQCs coated with 10.0 µg of TX-100 exposed to phenols.
A-the indicator sensor, B-the referent sensor
T_I-the first breakpoint or the penetration point, T_{II}-the second breakpoint
a) ACC impregnated with cooper tartarate, b) ACC impregnated with formaline, c) ACC impregnated with TX-100, d) ACC impregnated with 4-AAP.

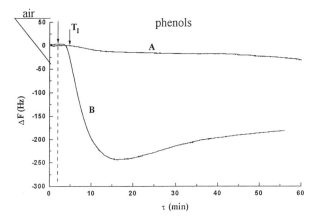

Figure 3. Breakthrough curve for phenols and the ACC impregnated with PVP. PQCs were coated with 10.0 µg of TX-100.
A- the indicator sensor, B- the referent sensor
T_I- the first breakpoint or the penetration point.

Conclusion

PQC sensor system can be successfully used for the evaluation and characterisation of an ACC in a continuous gas flow system. For the detection of phenols the best feature as the crystal coating exhibited Triton X-100 (10μg). For the removal and separation of phenols from the air the best sorption material was the ACC impregnated with poly(vinyl pyrrolidone). Protection time was more than 60 min.

Acknowledgement

Author gratefully acknowledges the support given by the Serbian Research Council

References

1) Sauerbrey, G. *Z.Phys.* **1959**, 155, 206
2) King Jr., W.H. *Anal Chem.* 1964, 36, 1735
3) McCallum, J.J. *Analyst* **1989**, 114, 1173
4) Rajaković, Lj.V. *J.Serb.Chem.Soc.* **1991**, 56,521
5) Rajaković, Lj.V.; Bastić, M.; Tunikova, S.; Bel'skih, N.; Korenman, Y.I. *Anal. Chim. Acta* **1995**, 318, 77
6) Thompson, M.; Kipling, A.; Dunkan-Hewitt, W.C., Rajaković, Lj.V. *Analyst* **1991**, 116, 881
7) Rajaković, Lj.V.; Ilić, M.R.; Jovanić, P.B.; Radošević, P.B. *Carbon* **1995**, 33, 10, 1433
8) Ilić, M.R.; Jovanić, P.B.; Radošević, P.B.; Rajaković, Lj.V. *Separation Sci.Technol.* 1995, 30, 2707
9) Ilić, M.R.; Jovanić, P.B.; Radošević, P.B.; Rajaković, Lj.V. *J.Serb.Chem.Soc.* **1995**, 60,507
10) (Rajaković, Lj.V., First paper in this book, in press)

Chapter 16

Trace-Moisture Sensor Based on the Electrolytic Technique

Naim Akmal[1] and Yining Zhang

Teledyne Electronic Technologies-Analytical Instruments, 16830 Chestnut Street, City of Industry, CA 91749

This paper describes the development of an electrolytic moisture sensor that can quickly detect trace levels of moisture in different gases such as N_2, Ar, He, H_2, O_2, and compressed air. This new moisture sensor uses a novel nanocompsite film as an electrolyte which is cast on an integrated array of electrodes made up of different metallic films as cathode and anode electrodes. A DC potential of 5 to 20 volts was applied across the electrodes. During the sensing process, moisture diffuses through the solid nanocomposite film and is electrolyzed into hydrogen and oxygen on the electrode surface. The produced current was measured and was used to determine the concentration of moisture in the sample gas. The response time for low ppb_v levels of moisture was found to be less than 30 min and the detection limit was 0.5 ppb_v. The sensor was found to be linear in the range of low ppb_v to 100 ppm_v levels of moisture. Also, the sensor was modified to work in hydrogen background avoiding any recombination phenomenon.

The measurement of moisture in gases is critical in chemical process industries, because of the awareness and control of water content in the feed streams. Reaction mixtures and product streams often determine the efficiency of a manufacturing process, quality of the product, and overall success of the process. The production of semiconductors and integrated circuits requires that the amount of moisture in the gases utilized during the fabrication process be quickly detected and measured to achieve better yields and reduce production costs. Presence of moisture in the semiconductor industry can cause particle formation and may result in electrical

[1]Current Address: Union Carbide Technical Center, 3200 Kanawha Turnpike, South Charleston, WV 25303.

©1998 American Chemical Society

property defect, that could adversely affect devices yield. In the polymer processing industry detection of moisture content in the monomer is very important to keep the reactors running at their maximum efficiency and control the desired molecular weight of the polymer . In corrosive and reactive gases, moisture contamination can severely damage gas lines, valves, pressure regulators, and mass-flow controllers. The moisture content and its measurement depends on the type of and nature of the process involved. Moisture levels below 100 ppb_v in inert gases are now commonplace, and several state-of-the-art facilities are operating below 10 ppb_v. Systems operating below 1 ppb_v have been reported (1). Measurement of moisture below 1 ppb_v level requires careful handling of the delivery system and an instrument that has a high signal to noise ratio.

Several techniques are available for detecting moisture in process gases (2-4). Such as, chilled mirror hygrometers, quartz crystal oscillators, capacitive sensors, and the highly accurate and expensive atmospheric ionization mass spectroscopy (APIMS) techniques. The application of the technique varies depending upon the process.

Chilled Mirror Hygrometers. Chilled mirror hygrometers measure the temperature at which moisture changes its phase on the surface of a cooled mirror (from gas to liquid or from gas to solid). At the frost point, ice on the mirror of the hygrometer is at equilibrium with water vapor in the gas. The detection is performed by scattering of light by the frost film. The light is generated by a LED source, and its intensity is measured by a detector. A change in the moisture content of the gas has the potential to change the thickness of the film that would in turn change the light-scattering intensity. The temperature of the mirror is varied to maintain a constant frost film and light-scattering intensity. The response time of such a device is very slow since it requires the moisture in the gas and the solid phase to be at equilibrium. In addition, this system is very bulky and involves a high maintenance cost. In order to observe a 10 ppb_v moisture in a process gas, the mirror is required to be chilled down to -104°C. In addition it can take up to 12-16 h to attain equilibrium at this low levels (<10 ppb_v) of moisture. Commercially available chilled mirror hygrometers are usually built around a bulky refrigeration system that restricts its movement and also requires maintenance at a high level. The National Institute of Standards and Technology (NIST) has rated this technique as a secondary standard for moisture concentration accuracy measurements (5).

Quartz Crystal Hygrometers. Moisture in gases can be measured by quartz crystal oscillators that are coated with highly hygroscopic resins. The sensing crystal is alternately exposed to the sample and reference gases. The reference gas is a dry gas, normally generated within the unit by passing the sample gas through molecular sieve based super dryers. The two streams are switched at a regular interval, usually every 30 s. Moisture in the sample gas is adsorbed by the hygroscopic coating, causing an increase in the mass of the crystal, which, in turn, decreases its oscillation frequency. The differences in crystal oscillation frequency between the reference and sample gases are used to compute the moisture in the sample gas. This type of moisture device is expensive and is not capable of detecting moisture levels below 5 ppb_v due to the poor signal to noise ratio. In order to an get accurate and reliable result, it is necessary to have a good and efficient moisture scrubbing system, and a reliable mass flow controller.

Capacitive Hygrometers. Capacitive sensors are made up of alumina or silicon dioxide materials sandwiched between two metal electrodes. The change in the moisture content in the surrounding medium changes the dielectric constant of the alumina, subsequently its capacitance. The capacitance change is then correlated to the moisture concentration in the gas stream. These sensors are being used to monitor moisture in inert gases, hydrocarbons, carbon dioxide, and other corrosive gases. Capacitive sensors are pressure sensitive and their response is extremely slow, since the moisture in the gas phase and moisture on the dielectric must reach an equilibrium before a stable signal is obtained. A constant drift in the output of the sensor requires frequent calibration.

Additional Techniques. Atmospheric pressure ionization mass spectroscopy (APIMS) is capable of detecting moisture in low ppt_v levels with confidence but due to its size and cost of ownership, (between US \$230,000 to 500,000) it is not always the best choice. The analyzers operation also requires specially trained technicians. Recently, the use of intracavity laser spectroscopy has been demonstrated to detect moisture in the gases (6).

Gas contaminant can also be determined by a Fourier Transform Infrared (FTIR) spectroscopy (6). FTIR spectroscopy measures the absorbance of infrared radiation by a contaminant in a gas sample as it passes through a sample cell. The change in absorbance is used to determine the concentration of contaminants. FTIR spectrometers are bulky in size and require calibration, purging with a dry gas and precise optical alignment for low levels of moisture detection.

Wei et al has recently reported a new in-line moisture analyzer based on the quartz crystal with a metal coating (7). This new moisture analyzer has a fast response time, is inexpensive and small, but cannot tolerate high levels of moisture due to its limited lifetime.

Electrolytic hygrometers rely on the absorption of water molecules by a highly hygroscopic material such as phosphorus pentoxide, which electrolyzes water and produces current directly proportional to the moisture content in the sample gas. Electrolytic hygrometers are being widely used in detecting trace levels of moisture in nitrogen and other inert gases. The original electrolytic hygrometer was invented by Keidel in 1958 (8). The Keidel cell made up of a body, which contains a pair of noble metal electrodes and is packed with a suitable water scavenger such as phosphorus pentoxide. A sample stream is introduced into the body and the moisture is retained by the hygroscopic scavenger. The retained moisture is then electrolyzed at the electrodes and the current required to electrolyze the moisture is measured. Additional work in this area was done by Goldsmith and Cox (9). They are convenient to use, inexpensive and offer very low levels of detection. However, the electrolytic moisture sensor has some major flaws. It is incapable of detecting moisture in hydrogen, oxygen and compressed gases due to the catalytic recombination of hydrogen and oxygen. Also, it cannot detect very low levels of moisture due to the nonelectrolytic current between the electrodes.

This work describes an improved electrolytic moisture sensor for inert gases, including hydrogen, oxygen, and environmental air, that is capable of sensing and signaling the moisture content in a few parts-per-billion level. Its stability, response and recovery times, and performance in corrosive gases has also been studied.

Experimental

Figure 1 shows the structure of the sensor. The sensor substrate with a dimension of 0.49 x 0.49 inches is made up of 99.6% pure alumina and was purchased form Coors Ceramics. An interdigitated array of noble metal electrodes were deposited on one side of the alumina substrate. This was accomplished by sputtering of the metal target. An adhesive layer of chromium was applied to the substrate before the application of the noble metal in order to achieve better adhesion of the metallic film to the substrate. The interelectrode gap range was approximately between 0.002 and 0.020 inches, depending upon the sensitivity required. The noble metals used were rhodium and platinum. One of the electrodes was deposited with a thin layer of pure gold, by electroplating in a gold salt solution. The thickness was controlled by the duration of the electroplating and the applied current to the electrode.

The sensor was cleaned by boiling it in a 30% hydrogen peroxide solution and then cleaning it in a dilute nitric acid solution. Finally the sensor chip was rinsed with deionized distilled water twice and dried under pure nitrogen.

Polymer Film. A 5% Nafion® solution was treated with a mixture of phosphoric and boric acids. Once prepared, the Nafion® solution was stored in a refrigerator to maintain its concentration. Before the application of the polymer film, the surface of the substrate was modified with a sol-gel to create a monolayer. A thin film of the above mentioned modified Nafion® solution was cast on the substrate by a spin coater. The rpm of the spin coater was optimized to get the desired thickness of the Nafion film.

The film coated substrate was treated at 120°C in a nitrogen atmosphere for 4-6 h in an oven. The cured sensor assembly was then mounted on four posts on a hermetic seal. Two of the posts were connected to the terminals of the electrodes; these acted as the anode and cathode. The hermetic seal was soldered to the top half of the sensor block. An electropolished cavity in the bottom half of the sensor block accommodated the electrode when the two halves were fitted together. A metal O-ring was used to seal the two halves to prevent adsorption/desorption of moisture from the sample gas or external leaks.

Sample System. The whole sensor block was attached to an electropolished stainless steel sample system and was maintained at 50±0.1°C using a thermal heater and a temperature controller. This sampling system is illustrated in Figure 2. Each major component of the sample system such as valves, flow controller, moisture scrubber and permeation tube was captured on the sample system using the screws and gasket technique. This ensured a leak free system. The calibration was performed using an internal water permeation tube that added moisture to the scrubbed sample gas to create a known concentration of moisture in the gas. Creation of moisture-free zero gas was achieved by installing a moisture scrubber upstream of the permeation tube. The scrubber consisted of a tube filled with the molecular sieve material. A mass flow controller was used upstream of the permeation tube to measure the rate of gas flowing by the permeation tube. Maintenance of a constant temperature and flow of the gas is necessary to generate an accurately known amount of the moisture in the system.

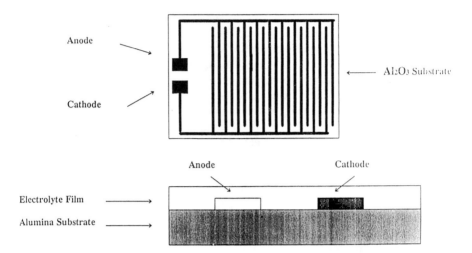

Anode

Cathode

Al₂O₃ Substrate

Anode

Cathode

Electrolyte Film

Alumina Substrate

Figure 1. Structure of the electrolytic moisture sensor.

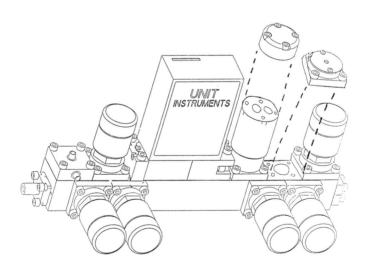

Figure 2. Sampling system of the sensor.

Results and Discussion

Operation of the Sample System. The flow schematic of the moisture analyzer is illustrated in Figure 3. The schematic has six air-actuated diaphragm valves. The purpose of these valve is to control delivery of the sample, span and zero gas through the sample system. These valves are driven by electrical solenoid valves. The sample system has four modes of operation:

Sample Mode:	PV2, PV4, PV5, and PV6 are on
Zero Mode:	PV1, PV3, PV4, PV5 and PV6 are on
Span Mode:	PV1, PV3, PV4 and PV6 are on
Purge Mode:	PV1, PV2, PV3, and PV5 are on

Mechanism. The moisture in the sample gas diffuses through the modified ion-exchange membrane and is subsequently electrolyzed to H_2 and O_2 by the applied potential. During electrolysis, electrons are released by the oxidation of the hydroxyl ion at the anode (Eqn. 2). The electrons released at the anode surface flow to the cathode surface where reduction of hydrogen ions occur (Eqn. 3). This current is measured and used to determine the concentration of moisture in the sample gas. The current generated during electrolysis is proportional to the partial pressure of water in the sample gas. Eqn. 4 is a simplified representation of the overall electrolytic process.

$$H_2O \dashrightarrow H^+ + OH^- \qquad\qquad (1)$$

$$2OH^- \dashrightarrow O_2 + H_2 + 2e^- \qquad (anode) \qquad (2)$$

$$2H^+ + 2e^- \dashrightarrow H_2 \qquad (cathode) \qquad (3)$$

$$2H_2O \dashrightarrow 2H_2 + O_2 \qquad\qquad (4)$$

In general, moisture diffuses through the moisture-sensitive layer to reach the electrode, and hence the determination of the concentration of moisture in the sample gas is dependent upon the partial pressure of water and not the flow rate.

In commercially available electrolytic hygrometers the electrolysis of water produces oxygen and hydrogen. These O_2 and H_2 products spontaneously recombine with the sample gas, if it is hydrogen, oxygen or a compressed air forming additional water molecules, due to the catalytic effect of the sensor electrodes. This recombination phenomenon results in a higher moisture reading than the actual moisture levels in the sample gas. The distortion of the moisture level due to this recombination effect becomes significant when the moisture level is <1 ppm_v in hydrogen, oxygen or compressed air. The recombination reaction can be shown in the following equation:

$$H_2O \text{ (sample gas)} \rightarrow H_2 + \frac{1}{2} O_2 \text{ (electrolysis product)} \qquad (5)$$

½ O_2 (electrolysis product) → H_2 (from sample gas) → H_2O (produced by recombination reaction)   (6)

This additional amount of moisture produces a false reading of moisture level in the sample gas.

Response in Nitrogen, Argon and Helium Gases. Figure 4 shows the response of the sensor for 250 ppb_v of moisture in a nitrogen gas. The moisture was generated from a moisture generation system and heat traced lines. The 90% response of the sensor for 250 ppb_v of moisture is less than 30 min. It takes approximately 2 h to attain equilibrium. This is due to the complete breakthrough of the moisture in the sample system. The recovery time has been found to be a little slower than the response time. This may be due to the surface condition of the sample system. Figures 5 and 6 show the response of 250 ppb_v of moisture in argon and helium gases respectively. The 90% response time in helium for 250 ppb_v of moisture is relatively slow. This is due to the difference in the molecular size of the sample gas and moisture.

Response in Hydrogen Gas. In the past, some success has been obtained in eliminating the recombination phenomenon by using two gold electrodes. To test this theory we used a pair of gold electrodes on an alumina substrate. Initial evaluation was successful, but after a few hours of usage the output dropped and finally there was no substantial change in the output of the sensor by varying the concentration of moisture in the sample gas. Migration of metal ions from one electrode to the other causes the output to continuously increase, until it reaches a point where the cell becomes electrically shorted. During this process the metal particles go into an oxidation process, i.e., they seems to be dissolved. This phenomenon is represented in Eqn. 7.

$$M \longrightarrow Mn^+ + ne^- \text{ (at anode)} \qquad (7)$$

$$Mn^+ \xrightarrow{migration} Mn^+ \text{ (to cathode)} \qquad (8)$$

$$Mn^+ + H_2 \longrightarrow M \text{ (at cathode)} + H^+ \qquad (9)$$

The ions migrate within the solid state electrolyte by the applied electrical field (Eqn. 8), and are then reduced by the hydrogen produced during the electrolysis process, as shown in Eqn. 9. This step occurs extremely fast in a hydrogen background. Finally, this metallic deposition (migration) fills the gap between the anode and cathode causing the electrode array to be electrically shorted. The migration effect depends upon the electrochemical oxidation potential of the anode materials and the value of the potential applied across the electrodes. The value of the applied potential across the electrodes cannot be reduced below a certain value, in order to keep a continuous dissociation of water. Different electrode materials have different catalytic effects on the recombination phenomena and have different migration effects upon the application of an applied potential.

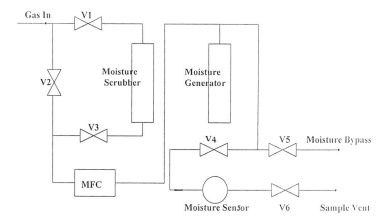

Figure 3. Flow schematic of the sensor system.

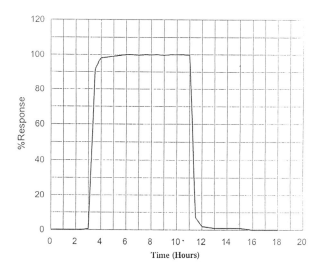

Figure 4. Response of the sensor for 250 ppb$_v$ of moisture in nitrogen gas.

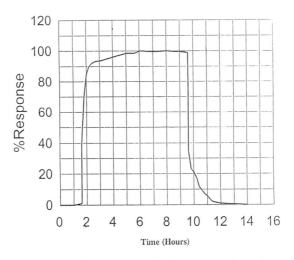

Figure 5. Response of the sensor for 250 ppb$_v$ of moisture in argon gas.

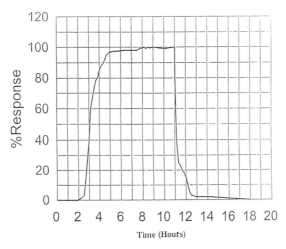

Figure 6. Response of the sensor for 250 ppb$_v$ of moisture in helium gas.

It has been found that a pair of platinum electrodes show the highest recombination effect in a hydrogen and oxygen background, but no migration effect was noted; whereas gold electrodes show the least recombination effect and maximum migration effect. Rhodium electrodes show very little recombination effect and virtually no migration effect. Therefore, it is reasonable to select gold as a cathode and rhodium as an anode for an electrolytic sensor to detect moisture in hydrogen and oxygen gases. Figure 7 shows a response of 250 ppb$_v$ of moisture in hydrogen gas. It is obvious that no recombination of moisture is taking place during the measurement. The recombination phenomenon has been eliminated by making one of the electrodes of rhodium and the other of gold.

Linearity. The electrolytic sensor has been found to be linear in all gases from ppb$_v$ range to ppm$_v$ ranges of moisture. Figure 8 and Figure 9 show the linearity of the moisture sensor in hydrogen, and helium gases respectively. Moisture for the linearity test was generated using a permeation device and maintaining it at a constant temperature and a flow rate. The sensor has been found to be linear for most of the ranges except around 400 ppb$_v$ levels of moisture . The slight non-linear behavior at this level may be due to the physical constraints of the sample system or due to the position of the sensor inside the sensor block. A few changes in the position of the sensor inside the block have produced better results.

Comparison with Available Techniques. Table I provides a list of various moisture detection systems in terms of cost and performance, and their comparison with the new polymer based electrolytic sensor. Most of the techniques cannot detect moisture below 10 ppb$_v$ level and their response and recovery times for low levels of moisture are in hours. The polymer film based electrolytic sensor can successfully detect moisture in single digit ppb$_v$ levels with a response time of a few minutes.

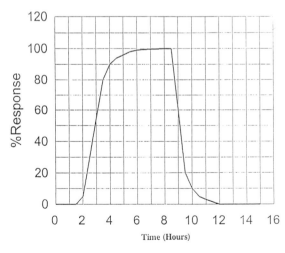

Figure 7. Response of the sensor for 250 ppb$_v$ of moisture in hydrogen gas.

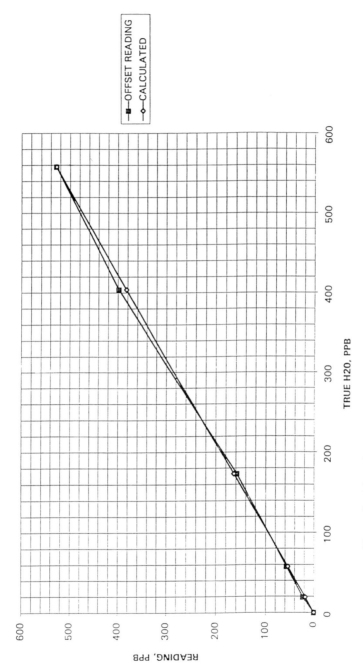

Figure 8. Linearity test of the sensor in nitrogen gas for 20 ppb$_v$ to 560 ppb$_v$ of moisture.

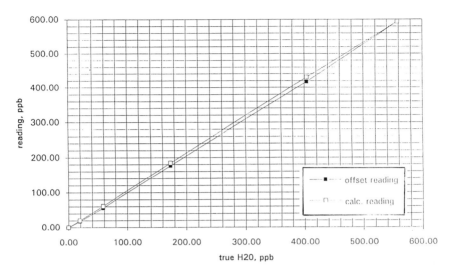

Figure 9. Linearity test of the sensor in hydrogen gas for 20 ppb$_v$ to 560 ppb$_v$ of moisture.

Table I Performance and Cost Comparison of Various Moisture Detection Systems

Characteristic	Chilled Mirror	Quartz Crystal	Capacitive	Electrolytic	APIMS	Modified Electrolytic
Response Time	2-3 h for ppb$_v$ levels of moisture	80% of value in 10 min at 10 ppb$_v$	2-3 h at ppb$_v$ levels	60-90 min at ppb$_v$ levels	Immediate	80% of value at 5 min at 10 ppb$_v$
Recovery Time	3-4 h when exposed to double digit moisture	10-15 min, when exposed to double digit ppb$_v$ levels	2-3 h at ppb$_v$ level	90-120 min at ppb$_v$ levels	Immediate	Less than 15 min when exposed to double digit ppb$_v$ levels
Limit of Detection	Single-digit ppb$_v$ level	Single-digit ppb$_v$	Double digit ppb$_v$	Double digit ppb$_v$	Sub-ppb$_v$	Sub-ppb$_v$
Working Range	0-1000 ppb$_v$ levels	1-1000 ppb$_v$ levels	100 ppb$_v$ - percent levels	ppb$_v$ to low ppm$_v$ levels	ppb$_v$ level	Single digit ppb$_v$ to 100 ppm$_v$ levels
Accuracy	1 ppb$_v$	10 ppb$_v$	20 ppb$_v$	10 ppb$_v$	10 ppt$_v$	0.5 ppb$_v$
Approximate Cost	$70,000	$39,000	$5,000	$23,000	$220,000 to $500,000	$28,000

Response to 0.50 ppb$_v$ of Moisture in Nitrogen Gas. Figure 10 shows the response of the sensor to a 0.5 ppb$_v$ moisture challenge. The peak-to-peak noise is less than 80 ppt$_v$, and the signal-to-noise ratio is approximately 9:1. This indicates that the sensitivity of the sensor is in the double digit ppt$_v$ region. The slow response time is due to the slow breakthrough of the moisture in the sample system. The system and the sample line were completely dried before introducing the 0.5 ppb$_v$ of moisture.

Long Term Stability. The output of the sensor registered a drop of approximately twenty percent during the first few weeks and therefore required frequent calibrations. After four weeks of continuous operation the output was stabilized, and the calibration was required only once a week. The initial twenty percent drift was due to the loss of water from the membrane and the substrate and is considered to be normal. The continued calibration that is required on a less frequent basis (weekly) is due to the conditioning of the membrane and the loss of conductivity. The effective life time of the sensor has been found to be roughly around 18-24 months. After this time period, the membrane may develop cracks leading to a loss in conductivity and eventually its moisture sensitivity. An added advantage of this sensor is that its exposure to high levels of humidity does not affect its performance, provided it is conditioned for a few days before being put into use for low levels of moisture (ppb$_v$) detection.

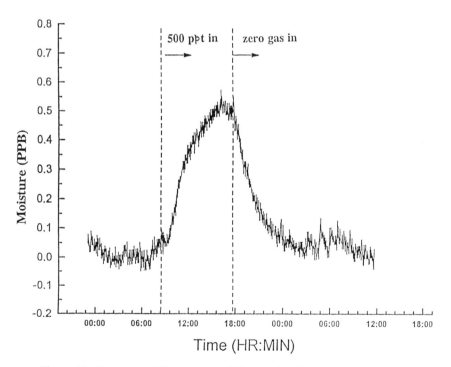

Figure 10. Response of the sensor to 500 ppt$_v$ of moisture in nitrogen gas.

Response in Carbon Dioxide and Corrosive Gases. An attempt was made to detect moisture in carbon dioxide gas. Initially the output was stable, but after a few hours of continuous operation it started to drop and finally no response was obtained. A careful observation of the sensor revealed that it had lost its conductivity in the carbon dioxide background. An attempt was also made to detect moisture in the corrosive gases such as chlorine, hydrochloric acid, etc. The sensor failed to respond. This is due to the nature of the corrosive gases which tend to destroy the polymer film, that is responsible for scavenging moisture from the sample gas.

Conclusion

In this paper a new generation of moisture sensors and their capabilities have been discussed. The new electrolytic based sensor, for the rapid detection of moisture in inert gases, will open the door to learn more about the process and requirements to control the moisture content, specially in the semi-conductor industry.

The sensor has been successfully demonstrated to perform in hydrogen and oxygen gases. A high signal to noise ratio allows us to successfully see a response to ppt_v levels of moisture in inert gases. Additional work is in progress to improve the response time of the sensor and its ability to detect moisture in corrosive gases. Effects of film thickness on the performance of the sensor are also being studied.

References

1. Yabumoto, N. et al. *Ultra-Clean Technology* **1990**, *1* (1), 13.
2. McAndrew, J. J.; Boucheron D. *Solid State Technology* **1992**, *35*, 55.
3. Bhadha, P. M. *Welding Journal* **1994**, *May*, 57.
4. Carroll, D. I.; Dzidic I.; Horning, E. C. *Applied Spectroscopy Reviews* **1981** *17(3)*, 337.
5. Hasegawa, S.*Int'l Sypm. on Moisture and Humidity*, ISA **1985**, Washington DC., 15.
6. Stallard, B. R.; Espinoza, L. H.; Niemczyk T. M. *Proceedings of the Institute of Environmental Sciences, 41st Annual Technical Meeting* **1995**, *41*, 1.
7. Wei, J.; Pillion, J., E.; King, S. M.; and Verlinden, M. *Micro* **1987**, *15* (2), 31.
8. Keidel, F. A. *Anal. Chem.* **1959**, *31*, 2043.
9. Goldsmith, P.; Cox, L. C. *Sci. Instrum.* **1967**, *44*, 29.

Chapter 17

Measurement of the Anesthetic Agent Halothane Using an Electrochemical-Based Sensor

Daren J. Caruana

Department of Biomedical Sciences, University of Malta, Msida, MSD 06, Malta

Anodic stripping chronoamperometry in acidic conditions is used to measure halothane at clinically relevant concentrations in the vapor phase (0 to 5%). The reduction of halothane in aqueous acidic solution is an irreversible process occurring at -0.15 V vs. SCE, that results in the deposition of a self-limiting layer on a gold electrode surface. This layer may be oxidized at +0.65 V vs. SCE. The influence of pH and chloride concentration on the reduction and oxidation processes are investigated and optimized in the presence of oxygen. The fabrication and performance of a thin film sandwich-type sensor is also described. The sensor is inexpensive to produce, has a fast response time (700 ms), and is uneffected by vapor flow rate.

Halothane ($CHClBrCF_3$) revolutionised inhalation anesthesia when it was introduced into clinical practice in 1956. Other halogenated hydrocarbons and ethers that have since been introduced as inhalation anesthetics have been modelled on halothane (1). Although popular as an anesthetic, the margin of safety is not wide and continuous breath by breath measurement of the anesthetic agent in the inspired and expired air is of great importance (2, 3).

In 1971 Severinghus (4) and others(5, 6) described that halothane interfered with the response from a Clark type oxygen electrode. The nature of this interference is now well understood and is attributed to the reduction of halothane at the electrode in the same potential region as oxygen (7). The concurrent reduction of oxygen and halothane made the measurement of either gas very difficult. Tebbutt and Hahn (8) managed, to a certain extent, to deconvolute the oxygen and halothane reduction waves by taking advantage of the difference in reduction potentials of the two gases. They reported that using this technique they were able to quantitatively measure oxygen and halothane at the same working electrode. Subsequent papers by Mount et al. (9, 10) described the reduction mechanism of halothane at a silver electrode in

188 ©1998 American Chemical Society

basic medium. The reduction is an ECEC reaction involving two electrons (see equations 1-4).

$$CHClBrCF_3 + e^- \rightarrow (CHClBrCF_3)^{\cdot-} \tag{1}$$

$$(CHClBrCF_3)^{\cdot-} \rightarrow (CHClCF_3)^{\cdot} + Br^- \tag{2}$$

$$(CHClCF_3)^{\cdot} + e^- \rightarrow (CHClCF_3)^- \tag{3}$$

$$(CHClCF_3)^- + H_2O \rightarrow CH_2ClCF_3 + OH^- \tag{4}$$

More recently, Langmaier and Samec (11) investigated the electrochemistry at various metal electrodes and in several solvents. In contrast with the study presented by Mount (9-10) they observed that the electrode material had a large effect on the reduction of halothane. The reaction proceeds via a specifically adsorbed intermediate.

$$(Ag...Br...CHClCF_3) \tag{5}$$

In the present study the conditions are manipulated to effect the strength of adsorption of this intermediate on gold.

Most reports on the electrochemistry of halothane have been concerned only with the reduction reaction occurring at approx. -0.3 V vs. SCE in aqueous alkaline solutions. In this study the reduction of halothane in aqueous acidic media containing chloride is described. Under these conditons reduction of halothane results in a strongly adsorbed species on the surface of the electrode (12). These products may be stripped off the electrode at high oxidation potentials. Anodic stripping voltammetry may be used to measure halothane concentration. At these high potentials there is no reduction of oxygen or any other gas that maybe present in the breath of patients. Therefore, this sensor is free from interference of other gases and the current response at this potential is dependent on the halothane concentration.

In this paper the electrochemistry of halothane reduction is described to determine the optimized conditions under which halothane may be measured most efficiently. The fabrication and performance of a thin layer sandwich type sensor, for the measurement of clinically relevant concentrations of halothane in the vapor phase is also described.

Experimental

Materials. All solutions were prepared using water from a Millipore MilliQ still (18 MΩ), and all glassware was soaked in 5% Decon (BDH, Poole, UK.) in distilled water overnight, washed thoroughly with MilliQ water and dried. Background buffer solution was prepared by mixing 0.1 mol dm^{-3} citric acid (BDH, Poole, UK.) with added 0.1 mol dm^{-3} sodium chloride (BDH, Poole, UK.) and 0.2 mol dm^{-3} disodium hydrogen orthophosphate (BDH, Poole, UK.) containing 0.1 mol dm^{-3} sodium chloride to obtain the desired pH, unless otherwise stated. Halothane (Fluothane, Zeneca, Macclesfield, UK.) was dissolved in methanol 1:1000 (v:v) dilution and this was added to a single compartment cell containing 10 ml of pH 4.0 buffer. The analogues of halothane, 2-bromo-1,1,1-trifluoroethane, 2-chloro-1,1,1-trifluoroethane and 1-bromo-1-chloroethane were purchased from Aldrich Chemical Co (Poole, UK). Nitrogen and oxygen (Medical grade, Multigas Malta Ltd.) were used to deoxygenate and oxygenate the solutions, respectively. Aliquots of methanol containing halothane were injected into the sealed two compartment electrochemical cell after purging for 15 minutes with the appropriate gas. The methanol itself did not contribute to the electrochemistry in any way. Halothane is supplied with 0.01% thymol as a stabiliser, this was removed by distillation under vacuum and used immediately. Unless otherwise stated, halothane was used as supplied.

An Abingdon-type vaporizer was used to deliver clinically relevant concentrations of halothane with nitrogen or oxygen as carrier gas. The gas/vapor mixture was passed through a bubbler containing water (to saturate the gas/vapor mixture to reduce evaporation of the cell solution) then purged through the test solution of pH 4.0 buffer solution (50 ml) for 15 minutes before each measurement was recorded.

Electrochemistry. All electrochemical experiments were carried out using a portable potentiostat (Oxford Electrodes) and the data was collected using a Lloyd Instruments PL2500 flat bed XY-t chart recorder. For solution electrochemistry of halothane a conventional three electrode set up was used comprising of a gold disc electrode of 0.126 cm^2 or 0.38 cm^2 diameter, a large surface area platinum counter electrode and a saturated calomel electrode (SCE) as reference. All cyclic voltammograms were recorded at room temperature (24 ±1°C). Cyclic voltammograms were digitized using a digitizing (Genius) tablet supported by software written by Dr. M. Balies, University of Bath.

A purpose built programmable potential step device with two potential settings, programmable delay and step duration for all anodic stripping chronoamperometry. The device circuitry was supported by software written in C. The potential step device was designed and built to specifications by A. Caruana, SGS Thomson, Malta Ltd. Current transients were recorded on a digital storage Oscilloscope model 310 (Nicolet) and transferred to PC. Sigmaplot (Jandel Scientific Software) was used in all data interpretation.

A potential step sequence used for the detection of halothane is shown in figure 1. For detection of halothane at a stationary gold disc electrode the following time/potential program was used: D_1= 100 ms at 0 V vs. SCE, D_2= 20 ms at P_1= -0.70 V, D_3= 100 ms at 0.0 V followed by a D_4= 250 ms pulse at P_2= +0.80 V vs. SCE, the sequence D_1 to D_4 being one event, the number of events (N) was 20 for each measurement.

Sensor element. Figure 2(a) shows a photograph of the sensor element, which consists of a two electrode system. The screen printed silver/silver chloride (Ag/AgCl) freference electrode (geometric area 2 mm x 5 mm) was covered with a

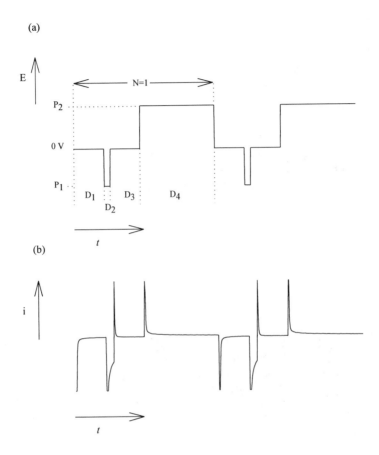

Figure 1. The potential step program for the measurment of halothane showing the potental time program for the detection of halothane (a), and the current transient (b), from a gold electrode in background pH 4.0 buffer.

piece of Whatman 1 filter paper 2.5 cm long and 4 mm thin which served as a wick to keep the reference electrode saturated with buffer. A piece of Whatman 5 filter paper geometric area 2 mm x 5 mm sputtered with 20 nm of gold using an Edwards Sputterer/Coater Model S150B and placed on top of the wick directly above the reference electrode with the gold surface facing up. The gold coating on the top filter paper is referred to as the working electrode. The buffer solution within the two layers of filter paper served as the electrolyte between the gold layer and the underlying reference electrode. The working electrode contact was achieved, using 5 cm of gold wire (0.5 mm diameter) mounted in a pulled pasteaur pipette, the end of the gold wire

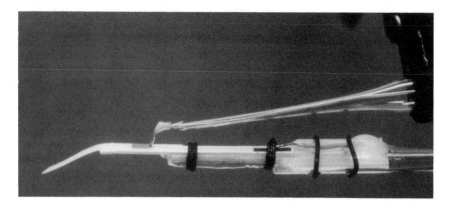

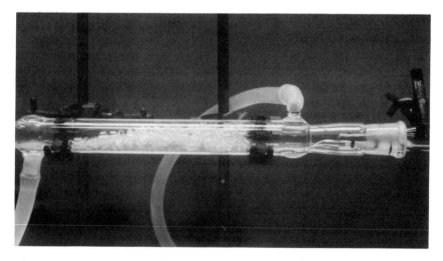

Figure 2. Photographs showing the sensor element with gold contact wire and filter paper wick (a), and the sensor positioned in the thermostated tube containing the wet glass turnings (b).

was sharpend to a point with a scalpel and gently placed on top of the gold coated filter paper, as shown in figure 2 (a). This configuration of electrodes ensured that there was a large surface area exposed to the gas phase with the minimum of cell resistance which could lead to iR drop.

The experimental set-up for sensor testing is shown in figure 2(b). The jacketed tube was connected to the circulation of the water bath and thermostated at 25 °C. Wet glass turnings were placed in the tube to moisten the halothane vapor delivered from the vaporizer. The gas entering the vaporizer was thermostated by placing 1 m of the tubing before the vaporizer in a water bath (Grants, Model Y6) at 30 °C. The total volume of the tube between the vaporizer outlet and the sensor element was 0.064 liters.

The same potential time program used to measure halothane at a gold electrode was again used for testing the sensor. For response time measurements the potential step sequence was changed to D_1= 0 ms, D_2= 20 ms at P_1= -0.655 V vs. Ag/AgCl, D_3 = 5 ms at 0.0 V followed by a D_4= 150 ms pulse at P_2= +0.750 V vs. Ag/AgCl and N= 30. The flow rate of the gases was measured using a Optiflow 520 digital flow meter (Humonics, USA).

Results and Discussion

The Electrochemistry of Halothane. Figure 3 (a) shows a cyclic voltammogram of 2 % halothane delivered from a vaporizer in nitrogen saturated pH 4.0 buffer between -0.5 V to 0.8 V vs. SCE. The cyclic voltammogram clearly shows a reduction at -0.15 V and a sharp stripping peak at +0.60 V, the latter is only observed on gold not on platinum or carbon. Scan rate analysis of both peaks showed that the reduction process is a solution borne species and the oxidation peak is from an adsorbed species produced during the reduction reaction (12). The products from the reduction form a self-limiting layer in the gold electrode surface which can be oxidised at +0.65 V vs. SCE.

The reduction peak is irreversible and resembles the reduction of halothane presented by Albery *et al.* (7) and others (5, 6, 9) at gold and silver electrodes in 0.1 mol dm^{-3} sodium hydroxide. The oxidation peak observed at +0.6 V vs. SCE has not been reported for halothane. The reduction of halogenated organics is dependent on solution conditions and electrode materal so reduction may occur by a number of distinct chemical pathways (13).

The Analogues of Halothane. As an initial attempt to identify the nature of the oxidation peak the electrochemistry of the analogues of halothane in pH 4.0 buffer was carried out, shown in figure 3 (b-d). Only one analogue 2-bromo-1-trifluoroethane, was reduced to any measurable extent, but all were found to have no anodic peaks. Halothane is supplied with 0.01 % thymol, which when removed by distillation leaves the electrochemistry unchanged. The electrochemistry of halothane is not trivial, and clearly involves specific adsorbtion of one or more products of the reduction reaction onto gold, possibly the intermediate shown in Equation 5 and described by Langmaier and Samec (11).

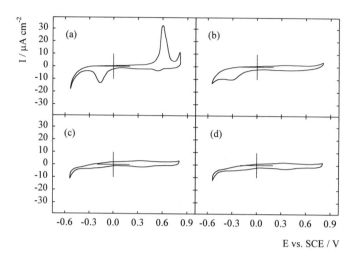

Figure 3. Shows cyclic voltammograms of halothane (2-bromo-2-chloro-1,1,1-trifluoroethane) (a), 2-bromo-1,1,1-trifluoroethane (b), 2-chloro-1,1,1-trifluoroethane (c), 1-bromo-1-chloroethane (d). At a stationary gold disc electrode at 50 mV s^{-1} between -0.5 and 0.8 V vs. SCE in background buffer.

The number of electrons transferred for the two Faradaic processes is not known, although the charge of the oxidation peak is exactly double that of the reduction peak, so one may postulate one electron transfer for the reduction and two for the oxidation. A series of experiments were carried out in order to understand greater the process involved. As stated previously the conditions for reduction of halogenated organics has a large effect on the reduction process and the resultant products. In this case an adsorption reaction is taking place which does not appear to occur as strongly in alkaline conditions. Thus in acidic aqueous solution the electrochemistry of halothane proceeds via an adsorbed species on the electrode surface similar to the process described by Langmaier and Samec (11).

pH Dependence. The pH dependence of the reaction was investigated by performing cyclic voltammetry at a clean gold stationary electrode in different pH buffered solutions. As shown in figure 4, there are no significant peak shifts observed in either the reduction or the oxidation peaks, suggesting that neither reaction involves protons. The oxidation peak does exhibit an increase when going to more acidic pHs. The reduction peak exhibits the reverse behavior as the pH decreases the peak current decreases.

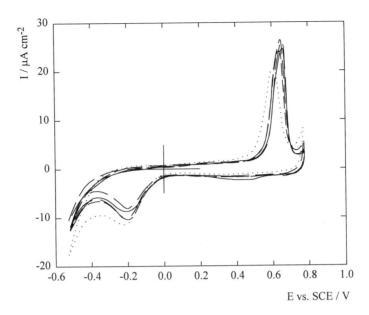

Figure 4. CVs of 4% halothane in pH 7 (·····), 6 (- - -), 5(– – –), 4(——) and 3(——) background buffer, scan rate of 50 mV s⁻¹, the 10th scan shown after the vapor was purged through the solution for 15 minutes.

The pH appears to effect directly the course of the reaction. In alkaline conditions the reaction occurs in a two electron reduction step. In acidic conditions the reaction appears to be a one electron reduction resulting with an adsorbed species followed by an oxidation process. There is further evidence to support this by measuring the oxidation peak width at half height. This decreases with decreasing pH from 90 mV at pH 7.0 to a value of 60 mV at pH 4.0. The value of the width at half peak height is dependent on the potential dependence of the adsorption energy (14). Essentially, the smaller the width at half peak height the stronger the species is adsorbed to the metal surface. The position of the peak is an indication of the free energy of adsorption which remains constant.

The pH is important to provide a suitable potential window. At alkaline pH the oxidation of gold occurs at sufficiently low potentials to mask the oxidation peak but under these conditions the reduction is more efficient. Generally, for halogenated hydrocarbons increased acidity decreases current efficiency because of the more extensive hydrogen evolution, although in some cases this is compensated for by the greater reduction facility as a result of faster protonation (13). However, in this case the pH has an effect on the strength of adsorption of the product of the reduction to the gold electrode, which in this instance is more dominant.

Chloride dependence. The oxidation process does not occur in the absence of chloride, figure 5 (a). in contrast the reduction occurs with or without chloride but if the products are not oxidised off the electrode further reduction cannot occur, the two peaks are mutually dependent.

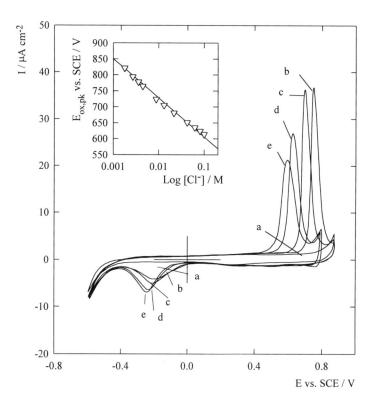

Figure 5. Cyclic voltammograms of 2 % halothane in pH 4 buffer containing [Cl⁻] = 0 mol dm⁻³ (a), 1.73 x 10⁻³ mol dm⁻³ (b), 3.54 x 10⁻³ mol dm⁻³ (c), 42.5 x 10⁻³ mol dm⁻³ (d), 100 x 10⁻³ mol dm⁻³ (e). Scan rate 50 mV s⁻¹, between -0.6 to 0.8 V vs. SCE at a stationary gold electrode.

The chloride ion may not directly be involved with electrochemical reaction, the halothane oxidation peak occurs as a 'prepeak' before gold chloride dissolution takes place (14). Figure 5 shows the CVs of halothane in sodium chloride over an increasing concentration range. There is a peak decrease and a negative potential shift at increasing chloride (Cl$^-$) concentrations. On closer analysis of the CVs the charge and the peak width at half peak height both remain unchanged with varying Cl$^-$ concentration. Interestingly, a plot of log [Cl$^-$] against peak potential (inset of figure 5) gives a linear plot with a 120 mV shift of the oxidation peak perdecade of Cl$^-$ concentration. There is no such effect observed when bromide (Br$^-$), fluoride (F$^-$) or iodide (I$^-$) are added.

A possible explanation for this is a catalytic or displacement reaction where gold chloride formation is required for the oxidation process to take place and which ultimately results in the oxidation of the products.

The electrochemistry of halothane appears to be complex in nature, however a good understanding has been acquired by studies done at a stationary electrode. Conditions have been optimized to enable the determination of halothane concentration by anodic stripping voltammetry, measuring the oxidation peak height or charge following the reduction at potentials lower than -0.3 V vs. SCE at the same electrode. The detection of halothane using cyclic voltammetry and potential step techniques and the influence of oxygen on the detection is presented below.

Halothane Detection using Anodic Stripping Voltammetry. Figure 6 shows the influence of peak height with lower negative potentials, the inset is a plot of negative potential against $I_{ox,pk}$ height. At a stationary electrode the plot is linear until the point where the electrode is saturated with adsorbed species, Γ_{sat}, which is proportional to electrode area. At coverages lower than Γ_{sat} the $I_{ox,pk}$ is proportional to halothane concentration. When halothane is delivered from a vaporizer, the concentration of halothane is relatively high and the lower potential limit needs to be raised to avoid reaching Γ_{sat}.

With an negative potential limit of -0.2 V vs. SCE, a plot of oxidation peak height ($I_{ox,pk}$) against halothane concentration delivered from a vaporizer is linear. When the same experiment was carried out in oxygen saturated solution, the $I_{ox,pk}$ against halothane concentration was linear, but the gradient of the line is roughly half that in oxygen free solution (figure 7 (a)). Furthermore, when low halothane concentrations were used in the range of 15 to 155 nmol dm^{-3} with a lower negative potential of -0.7 V vs. SCE there was no difference in the gradient between oxygen and nitrogen saturated solutions.

The interference of oxygen on the magnitude of the $I_{ox,pk}$ is dependent on hydrogen evolution. This was shown by potential stepping to different negative potentials followed by scanning from 0.0 to +0.8 V to measure the $I_{ox,pk}$ in nitrogen and oxygen saturated buffer at pH 4.0 and 7.0, shown in figure 8. For nitrogen saturated buffer at pH 4.0, $I_{ox,pk}$ increases with increasingly negative potential as shown in figure 6, followed by a decrease in current beyond -0.6 V vs. SCE which is attributed to increased hydrogen evolution. A similar behavior is observed in oxygen saturated buffer although there is a significant reduction in current before the onset of

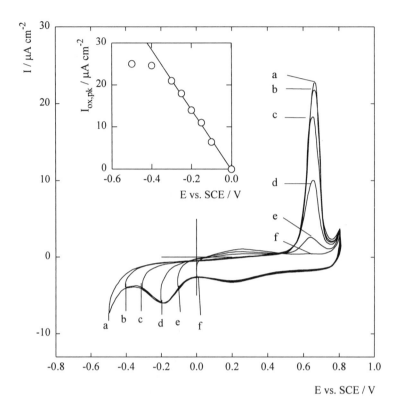

Figure 6. Showing CVs of 4 % halothane at scan rate 50 mV s⁻¹ at a gold electrode between +0.8 V and -0.5 V (a), -0.4 V (b), -0.3 V (c), -0.2 V (d), -0.1 V (e), -0.0 V (f) vs. SCE. The cycle has started at 0.0 V with a negative potential direction. In pH 4.0 background buffer soltuion. The inset shown is a plot of $I_{ox,pk}$ height against negative potential limit in the cyclic voltammogram.

hydrogen evolution. However in the hydrogen evolution region when a decrease in current is observed the current is the same as in nitrogen saturated buffer. The experiment was repeated in pH 7.0 buffer and again a similar effect was observed. However, the potential at which the current started to decrease was -0.9 to -1.0 V for nitrogen and -1.0 V for oxygen. There is also a 60 mV shift per pH unit which is as expected for a one electron process of hydrogen evolution in acidic solution: (15)

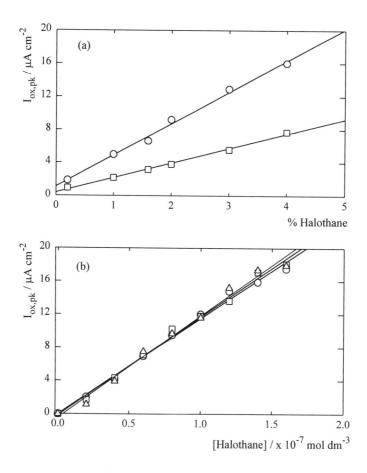

Figure 7. Showing (a) calibration plots of halothane measured by cyclic voltammetry between -0.2 to 0.8 V vs. SCE at 50 mV s⁻¹ delivered from a vaporizer in background buffer and (b) halothane dissolved in methanol between 20 and 155 nmol dm⁻³ and cycled between -0.7 to 0.8 V vs. SCE at 50 mV s⁻¹ added to background buffer solution in a sealed electrochemical cell. (circle) nitrogen, (triangle up) air and (square) oxygen saturated buffer.

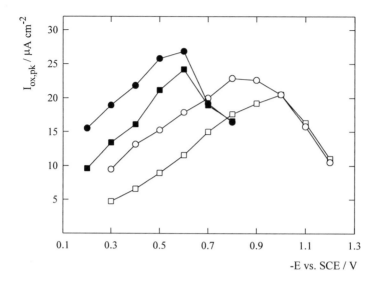

Figure 8. Showing a plot of $I_{ox,pk}$ against negative potential to which the potential was stepped in background buffer saturated with 4% halothane in pH 7.0 (hollow symbols) and 4.0 (filled symbols) buffer. The potentail step of duration 20 ms followed by potential scan from 0.0 to 0.8 V vs. SCE at a scan rate of 50 mV s^{-1} in nitrogen (circle) and oxygen (square) saturated buffer.

$$2H^+ + 2e^- \rightarrow H_2 \qquad\qquad\qquad (6)$$

The decrease in $I_{ox,pk}$ can be accounted for by an increase in the hydrogen evolution reaction, which reduces the reduction efficiency of halothane. This shows that oxygen reduction appears to be directly effecting either the reduction of halothane and/or the deposition of the adsorbed species from the reduction reaction. A strong possibility for the interference could be of a chemical nature, perhaps hydroxide ion attack of the radical formed from the first electrochemical step in the reduction of halothane. At alkaline pH there is more OH$^-$ produced from the reduction of oxygen consequently there is more interference on halothane reduction, although as yet there is no firm evidence to support this. Essentially, the interference of oxygen is eliminated by stepping to -0.7 V at the start of hydrogen evolution pH 4.0 buffer.

Halothane Detection using Anodic Stripping Chronoamperometry. An experiment was carried out to determine the optimum duration of the deposition step at -0.7 V vs. SCE. A plot of step duration against $I_{ox,pk}$ is shown in figure 9.

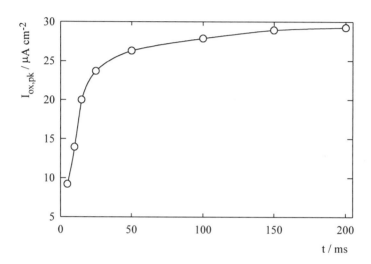

Figure 9. Showing step duration versus $I_{ox,pk}$ in 4% halothane delivered from a vaporizer with nitrogen as the carrier gas. The potential was stepped to -0.7 V vs. SCE, then scanned from 0 to 0.8 V at a scan rate of 50 mV s^{-1} in pH 4.0 buffer.

The current starts to saturate beyond 30 ms, where total coverage (Γ_{sat}) is reached. a partial coverage is required The optimum step duration for measurement of halothane using anodic stripping chronoamperometry is 20 ms, which corresponds to the linear region of the graph well before saturation starts to occur.

Pulse deposition followed by anodic stripping by voltmmetry is a good technique to provide information on the optimal conditions required for detection of halothane. However, for an analytical device to measure halothane it will be more convenient to use a double pulse method of detection anodic stripping chronoamperometry. With the proposed protocol having the first pulse to -0.7 V for 20 ms, resulting in the deposition of the adsorbed species and subsequently a pulse to +0.8 V to strip the oxidized products off the electrode. This was done first on a stationary gold disc electrode, with the resulting transients from the stripping step to +0.8 V are shown in figure 10 in nitrogen (a) and oxygen (b) saturated buffer.

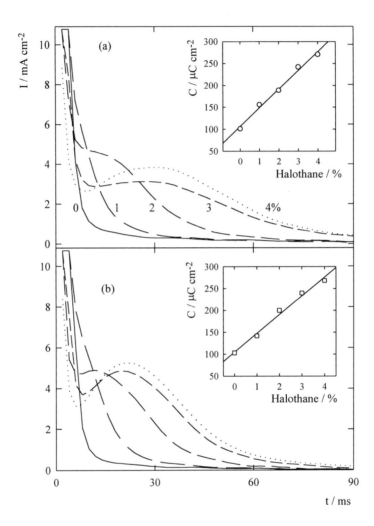

Figure 10. Transients for halothane (% shown in (a)) delivered from a vaporizer with nitrogen (a) and oxygen (b) saturated background buffer. Potential stepped to -0.7 V for 20 ms then to -0.8 V vs. SCE for 250 ms (90 ms shown), for 20 cycles (N). The 8th successive stripping transient is presented. The insets show the total integrated charge vs. halothane concentrations. Gradients are 46.9 and 46.2 μC cm^{-2} /(% halothane) for nitrogen and oxygen respectively.

The transients shown in figure 10 in nitrogen and oxygen saturated solution are similar, there is a peak where the material situated on the surface of the electrode is oxidized at once. At higher halothane concentration the peak takes longer to appear. This is due to the non- faradiac charging of the electrode, where at higher coverage, Γ, the double layer takes longer to charge up and delays the appearance of the peak. This is true for both nitrogen and oxygen saturated solutions.

Integration of the transients gives the total material that is oxidized plus the non-faradaic current. The linear plots of charge for the oxidation step against halothane concentration are shown in the insets of figure 10 (a) and (b) respectively. The gradients for the lines given in the figure legend are similar thus showing that detection of halothane in the way described here is not affected by oxygen.

A method has been presented here by which halothane may be detected free from the interference of oxygen using a double potential step technique. The rest of the paper will be describing the development and performance of a sensor which may be used to measure halothane in the gas/vapor phase.

Sensor Device Performance. There are certian practical conciderations when using a thin layer electrode such as the one used in this sensor. If the potential limits are extended to where gold chloride dissolution occurs, the current rapidly diminishes as the sputtered gold around the gold contact wire is dissolved and the electrical contact is broken to the rest of the gold. Furthermore, the cyclic voltammetry of the gold coated filter paper, figure 11, shows a broad redox couple at +0.25 V vs. Ag/AgCl. This process is probably due to the inherent nature of the gold surface produced by sputtering. From scan rate analysis the couple was found to be a surface process, present from the first cycle and does not significantly change in intensity with successive cycles upto 1 h. Interestingly when halothane is reduced on the gold coated filter paper there is a decrease of peak height of the surface process, probably due to some inhibiting phenomenon of the halothane reduction, the nature of which was not studied further.

The transient for the stripping step shown in figure 12 is not the same as on a solid gold electrode, there is no peak observed. There are two Faradaic processes clearly visible, the first at short time scales up to 30 ms and the second occurs beyond 30 ms. The first process is associated with the faradaic process shown in figure 11 occuring at 0.25 V vs. Ag/AgCl. Both processes are effected by the concentration of halothane. As the concentration of halothane is increased the Faradaic charge from the adsorbed process decreases linearly.

However, the process of interest is the oxidation of halothane which occurs after 30 ms. The integrated charge between 30 and 150 ms is proportional to halothane concentration. A plot of the charge between 30 and 150 ms against halothane concentration is linear and totally free of interference from oxygen as shown in the inset in figure 12.

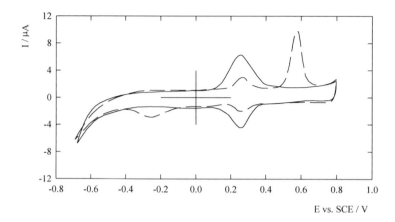

Figure 11. Cyclic Voltammograms of gold coated filter paper in the absence (——) and presence (----) of 2 % halothane with nitrogen as the carrier gas. Cycled at 50 mV s⁻¹ between -0.7 to +0.8 V vs. Ag/AgCl, the filter paper was soaked in pH 4 background buffer.

Response Time.   One of the more important technical considerations for breath-by-breath monitoring during anesthesia is the response time. The gold working electrode positioned in such an exposed orentation ensures the optimun design for a fast responding sensor for halothane in the vapor phase.

The response time is dependent on the thickness of the buffer layer at the surface of the gold, and the partitioning of halothane between the buffer layer and the gas phase. Figure 13 shows a detailed schematic of the sensor device and the thin film of buffer on the surface of the gold working electrode where the electrochemistry takes place.

Provided the buffer layer is sufficiently thin then there should be no dependence on mass transport and, thus, should be unaffected by flow rate. The detection of halothane is not   done continuously, sampling occurs at periodic time intervals by a potential step to -0.655 V vs. Ag/AgCl. The electrochemistry of the halothane is occurring in the thin layer of buffer on the gold surface. The diffusion layer of the deposition step can be estimated using the following equation, (16)

$$X_d = \sqrt{2Dt}$$

(7),

where D is the diffusion coefficient of halothane and t is the potential step duration which in this case is 20 x10^{-3} s. Langmaier et $al.$ (11) determined the diffusion coefficient to be 4.8 x 10^{-6} cm^2 s⁻¹ in aqueous solution, using equation 7, giving a diffusion layer of approximately $4.6 \pm 0.25 \ \mu m$. Consequently, as long as the layer of buffer on the surface of the gold is approximately the thickness of the diffusion layer, then the electrochemical response is dependent on halothane concentration and the partitioning of halothane between the gas phase and the buffer. It was found that provided the sensor element was not emersed in buffer and the halothane vapor was kept saturated with water the response is independent of flow rate. There was no significant change in halothane response at flow rates from $0.5 \ l \ min^{-1}$ to $1.5 \ l \ min^{-1}$, as shown in figure 14, the maximum plateaus charge at different flow rates is the same.

To ensure efficient functioning of the sensor the halothane vapor was kept wet to maintain this layer. When the glass turnings were removed from the jacketed tube the response of the electrode to halothane was reduced as the surface of the electrode becomes dry. If the glass turnings are maintained wet the sensor may be operational

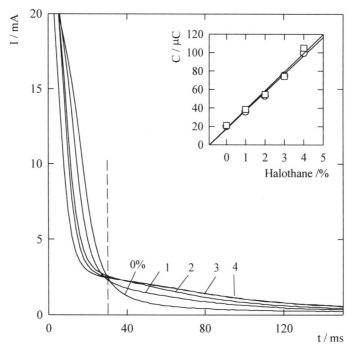

Figure 12. The transients obtained from the gold coated filter paper in a two electrode system for different concentrations of halothane with nitrogen as the carrier gas. The inset is a graph of the integration of the charge from 30 ms to 150 ms against the halothane concentration in nitrogen (circle) and oxygen (square) environment. The potential program described in the experimental section was used with N=20. The transient presented is the 8th successive step there is no gradual decrease in charge on successive potential step events.

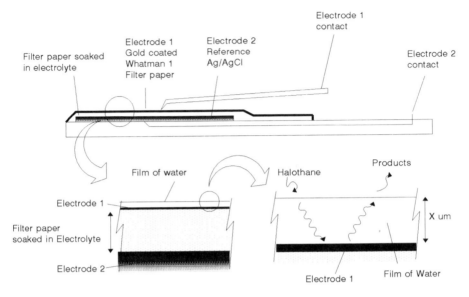

Figure 13. Schematic diagram and expanded views of the elecrode configuration in the sensor element used for the measurement of halothane in the vapor phase.

for more than 2 h without any detectible loss of response. By maintaining the vapor saturated with water the thin layer of buffer on the surface is maintained at the correct thickness for efficient functioning of the sensor.

The response time was tested using different vapor flow rates. The time resolution of this experiment is proportional to the sampling time which was 175 ms. Figure 14 shows the response from a single sensor at different vapor flow rates. There was an inital lag time which is the time taken for the vapor to reach the sensing element, followed by a rise until saturation occurs. The rise in response took approximately 700 ms at each flow rate to reach a plateau. If the buffer layer thickness on the gold surface is changed the response time and plateau charge are different.

The 'refresh` time of the sensor element was also tested. The time taken for the sensor response to fall from 4% to 0% halothane was slower to those obtained for the rise time, although at prolonged periods with high halothane concentration a slower refresh time of up to 1 s was observed.

Conclusions

The electrochemistry of halothane in aqueous solution on a gold electrode is more complex than has previously been proposed (6-10). The formation of a self limiting layer from the reduction reaction in acidic aqueous conditions is unique to halothane. Although the exact mechanism of the reduction has yet to be fully understood, the report presented here has shown that under defined conditions, halothane may be measured electrochemically in the presence of oxygen.

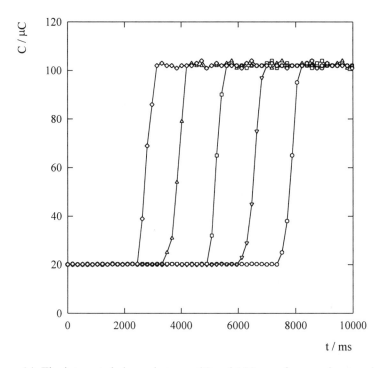

Figure 14. The integrated charge between 30 and 150 ms of successive transients against time at vapor flow rates 0.5 (circle), 0.75 (triangle down), 1.0 (square), 1.25 (triangle up), and 1.5 (diamond) l min^{-1}. The halothane concentration was increased from 0% to 4%. the potential step program used is described in the experimental section, N=50. All measurments were recorded with one sensor element with nitrogen as the carrier gas.

In alkaline aqueous solution the reduction of halothane is masked by the concurrent reduction of oxygen which makes the measurement of halothane in the presence of oxygen difficult. In acidic conditions the reduction of halothane results in an adsorbed self limiting layer which is oxidized at +0.65 V vs. SCE. The strength of adsorption increases with decreasing pH. Chloride is essential for the oxidation peak to occur, which may be a 'prepeak` of gold chloride dissolution.

Anodic stripping chronoamperometry, involving a deposition step followed by a stripping step has, been shown to be suitable for the measurement of halothane. If the deposition step is at a low enough potential to include hydrogen evolution the response to halothane is totally uneffected by oxygen.

A thin layer gold sensor element was fabricated and tested to measure clinically significant concentrations of halothane delivered from a vaporizer. The sensor response was stable for over two hours of continuous use, showed no dependence on vapor flow rate and the response time was calculated to be ca. 700 ms.

The sensor works best in acidic pH which is highly desirable as the exhaled carbon dioxide in the breath of the patient will maintain a low pH. The sensitivity of the sensor can be adjusted by simply extending the duration of the deposition step, or by employing coatings on the surface of the gold layer to facilitate more halothane to partition close to the electrode surface. Such a small sensor may be used to measure halothane in the surgical environment.

Acknowledgments

The author wishes to thank, Profs. J.V. Bannister and W.H. Bannister for support of this work. Mr. J. Giglio and Dr. H-H. Frey, Dr. M. Falzon and Dr. C.E.W. Hahn for helpful discussions. St. James Clinic Malta for the loan of the halothane vaporizer used throughout this work and SGS Thomson Malta Ltd. with assistance in preparing the gold coated filter paper.

Literature Cited

1. McIntyre, A. R., *Anesthesiology* **1959**, 20, **409**.
2. Dobkin, A. B., Development of new Volatile Inhalation Anaesthetics, North Holland Biomedical Press, Amsterdam, 1979, 78.
3. Graventein, J.S.; Paulus, D.A.; Hayes, T.J., Gas Monitoring in Clinical Practice, 2nd Edition, Butterworth-Heinemann: Newton, USA,1995,7.
4. Severinghus, J.W.; Weiskopf, R.B.; Nishimura, M., *J.Appl. Physiol.* **1971**, 31, 640.
5. Dent, J.G.; Netter, K.J; Br. J. Anaesth., **1976**, 48, 195,
6. Norden, A.G.W.; Flynn, F.V. *Clinica Chimica Acta,* **1979**, 99, 229.
7. Albery, W.J.; Hahn, C.E.W.; Brooks, W.N., *Br. J. Anaesth.* **1981**, 53, 447.
8. Tebbutt, P. and Hahn, C.E.W., *J.Electroanal. Chem.*, **1989**, 261, 205.
9. Mount, A.R.; Appleton, M.S.; Albery, W.J.; Clark, D.; Hahn, C.E.W., *J.Electroanal. Chem.*, **1992**, 334, 155.
10. Mount, A.R.; Appleton, M.S.; Albery, W.J.; Clark, D.; Hahn, C.E.W., *J.Electroanal. Chem.*, **1992**, 340, 287.
11. Landmaier, J.; Samec, Z. *J.Electroanal. Chem.* **1996**, *402,*,107.
12 Caruana, D. J.; Giglio, J. *J. Chem. Soc. Faraday Trans.* **1996**, 92, 3669.
13. Feoktistov, L.G., in *Organic Electrochemistry*; Editor, Baizer, M.M.; Lund, H., 2nd ed, (1983) 259
14. Wopschall, R. H.; Shain, I. *Anal. Chem.* **1967**, *39,* 1514.
15. Southampton Electrochemistry Group; *Instrumental Methods in Electrochemistry*; Ellis Horwood series in Physical Chemistry; Ellis Horwood: Southampton, 1993, 233.
16. Bard, A.J.; Faulkner, L.R., *Electrochemical Methods;* John Wiley & Sons, Inc.: 1980, 129.

ION-SELECTIVE ELECTRODES AND POLYMER-BASED SENSORS AND BIOSENSORS

Chapter 18

Sensors Based on Organic Conducting-Polymer Electrodes

Ahmed Galal[1], Nada F. Atta[1], and Harry B. Mark, Jr.[2]

[1]Department of Chemistry, Faculty of Science, University of Cairo, Giza, Egypt
[2]Department of Chemistry, University of Cincinnati, Cincinnati OH 45221–0172

Studies of the electrochemical behavior of electrodes modified with polymeric materials are one of the most rapidly growing and advancing areas of modern electrochemistry [1]. This is due to both the stability of these films produced electrochemically and to their interesting electrochromic and conducting properties. The selective response of this class of polymers towards dissolved ions has made them useful in various applications as a new "generation" of ion and molecular sensors. Adams and coworkers reported the original use of voltametric carbon paste electrodes to detect fluctuations in the concentration of compounds in brain extra-cellular fluid [2]. Much of the effort invested recently in the development of selective sensors for in vivo applications has been based on the elimination of the ascorbate contribution to the recorded signal, using anion repelling agents such as stearic acid or Nafion [3]. Electroactive polymeric films have acquired wide popularity since they are easy to generate at the surface of the electrode when compared to the monolayer approach [4]. Moreover, the relatively increased number of active sites throughout the polymer film rendered the electrochemical processes at its surface more pronounced when compared to monolayer-modified ones. During the course of development in the area of electrochemical sensors the strive for improvements in the stability, selectivity, and scope of such sensors imposed the real challenges towards the "successful" analysis of clinical and environmental samples. On the other hand, electrochemical methods of detection following chromatographic separation represent a cornerstone in the determination of catecholamines [5] and other ionic species.

The electrode materials most commonly used in amperometric systems are glassy carbon, platinum and gold electrodes [6]. However, several limitations were reported in the application of ca mercury electrodes in flowing solutions such as their poor mechanical stability and the dissolution of mercury at low potentials which resulted in their inadequacy in the detection of organic and biological compounds. Moreover, the performance of glassy carbon, pyrolytic graphite, platinum and gold electrodes depends mainly on the method and quality

©1998 American Chemical Society

of surface polishing and pretreatment. Glassy carbon electrodes were found to be sensitive to high current densities and to aging [7]. The use of carbon paste electrodes is also limited when organic solvents are implemented and the pastes are not resistant toward mechanical damage exerted by relatively high flow rate of solutions. Changes caused to the electrode surface due to the adsorption of the analytes, oxidative or reductive byproducts results in the commonly known as the electrode surface fouling. This phenomena represents a substantial challenge to the electrochemical detection of organic compounds. In previous work [8], we examined the electrochemical behavior of a large number of biologically important compounds at a P3MT electrode. The electrocatalytic property of the polymer films, which is independent of the nature of the substrate, was not yet demonstrated. In this work we examine and compare the electrochemical behavior of some neurotransmitters at different conducting polymer electrodes. Moreover, the advantages of using the polymer electrodes for the HPLC/EC separation and detection of some neurotransmitters will be presented. The polymer electrodes showed promising antifouling characters over the conventionally used glassy carbon and platinum electrodes. Examples will be provided for the detection of biological compounds in the presence of albumin, gelatin and Triton 100X. In a previous work [9], we reported the construction and properties of a polymer-based iodide-selective electrode. This electrode suffered from short life times, up to a maximum of fifteen days, but possessed a relatively high selectivity coefficients towards most interfering anions. Also, some technical data, such as film thickness, working temperature range, etc. were not reported. In this study, we also introduce the design of a new cap for the storage of this electrode which resulted in extending its life time to more than a year. The effect of changing the type of the conducting polymer used in the fabrication of this type of I^--selective electrode is given and the responses are compared. Moreover, the efforts towards the fabrication of electrode sensors for other anions are discussed.

EXPERIMENTAL

The fabrication of the ion-selective electrode has been described previously [9]. Electrochemical polymerizations were carried out under constant applied potential of + 1.7 V at a Pt, GC or graphite substrates as mentioned elsewhere [10]. All the potentials in the polymerization and other measurements were referenced to an Ag/AgCl (3M KCl) electrode. In the case of high performance liquid chromatography (HPLC) or flow injection analysis (FIA) followed by electrochemical detection (EC), the current signal responses were measured for an injected analyte sample. In this case the potential of a platinum or a polymer coated platinum electrodes imbedded in a thin-layer flow cell is monitored. The FIA and HPLC systems were followed by an electrochemical detection unit as was described in a previous work[10]. The applied potential used for the amperometric mode of detection was +0.55 V, unless otherwise stated, and the sensitivity range was 10 nA/scale. The mobile phase was either 0.2 M phosphate

buffer/0.05 M NaCl, pH = 6.9 or 0.15 M o-chloroacetic buffer, pH = 3.0
containing 0.8 mM sodium octyl sulfate. The mobile phase was kept under argon
atmosphere and was run isocratically at room temperature at a 1.00 ml/min flow
rate. Samples and standards in a volume of 20 µl were injected into the mobile
phase by using an Altex Model 100A (Altex Scientific Inc., Berkeley, CA) dual
reciprocating pump and an injection port Model 7120 (Rheodyne, Colati, CA).

All electrolytes and solvents were used as supplied. The polymer films
were undoped by subjecting the working electrode to an applied potential of
-0.2V in the same synthesis solution for 45 minutes, where the current decreases
to a 0.001 µA value within 5 minutes. The electrode was then removed from the
synthesis cell, rinsed with acetonitrile, alcohol and water. The electrode was
"activated" for subsequent use as a selective electrode by one of the following
methods; (i) the electrode was immersed in a cell containing aqueous solutions of
0.1 M KI, or KBr, or KCl, or K_2SO_3, and was subjected to a positive potential
(determined by the oxidation potential of the anion of interest) in order to redope
the film with the anion of interest. The electrode was then rinsed thoroughly with
water. (ii) The electrode was alternatively "activated" by exposing its surface to
one of the following gases: I_2, Br_2, Cl_2, or SO_2; the electrode was then rinsed
with water and dried; or (iii) the "activation" was a combination of the
electrochemical doping and the chemical treatment mentioned in (i) and (ii),
respectively. The variation of the method of "activation" resulted in large
differences in the potentiometric performance of the electrode. Potential
measurements were made as we described earlier[9]. These measurements were
normally carried out at 23 ± 0.5°C (except in the temperature range response
measurements). Cyclic and double potential step experiments were performed
using a BAS-100 instrument (BAS, Inc., West Lafayette, IN.).

RESULTS AND DISCUSSION

Conducting polymer ion-selective electrode.
The effect of the doping potential on the response of the conducting polymer
ion-selective electrode (CPISE) was studied for the poly(3-methylthiophene)
(PMTCPISE) and the poly(aniline) (PACPISE) films. Figures l(A) and l(B).
show the effect of the doping potential on the response of 1000 Å-thick film
electrodes of (PMT) and (PA), respectively. The effect of the applied potential
for the doping step was found to depend on the nature of the doping ion and the
type of the film under investigation. It is evident that the films doped at a
potential ±1.2 V did not produce an acceptable linear dynamic calibration range.
The films doped at potentials of +0.70, +0.80, or +1.00 V, on the other hand,
produced calibration curves with a larger linear dynamic range (10^{-1} - 10^{-7} M)
and with correlation coefficients of 0.998, 0.999, and 0.998, respectively.
Although the correlation coefficients of the calibration curves obtained for the
electrode doped at +0.70, +0.80, and +1.00 V were comparable, the slopes
obtained were different. For that doped at +0.80 V, the slope was 58.54 mV per
decade. This is close to the expected value of 58.7 mV per decade (at 23°C).

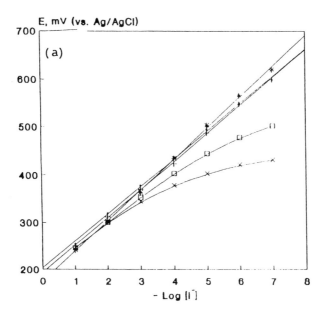

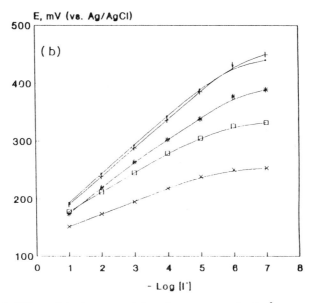

Figure 1 Effect of doping potential on the response of 1000Å-thick film electrode of (a) PMT, (b) PA: 0.70 V (-•-), 0.80 V (+), 1.00 V (*) 1.20 V (□), 1.50 V (X).

However, for the other two doping potentials the slopes were 57.42 and 64.54 mV per decade, respectively. The slope decreases significantly for films doped at potentials > +1.00 V. This phenomenon is probably a result of oxidative effects caused by the iodine (I_2 gas) evolution at the film surface at the higher doping potentials. Similar behavior was also observed in the case of the PACPISE. However, the slope of the calibration curves were 20, 40, 48, and 46 mV/decade, with a noticeable deviation from linearity at concentrations $\geq 10^{-6}$M of iodide The magnitude of the doping potential at which the PACPISE response decreases occurs at less positive potentials than that of the PMTCPISE. A possible explanation for this behavior is attributed to the less positive oxidation potential for iodide at the PA film (ca. 682 mV, as measured from the DPV experiments). In general PMT electrodes had better analytical performance compared to PA and poly(N-methylpyrrole) (PNMP) as shown in Table I.

The effect of film thickness on the potentiometric response of the modified PMTPCISE and PACPISE electrodes was also studied. For the PMTCPISE, the correlation coefficients and slope values are given in Table I for two different concentration ranges and indicate that all five electrodes (films grown at a constant deposition potential of +0.8 V, and with apparent thicknesses corresponding to 250 Å, 500 Å, 1000 Å, 1500 Å, and 2000 Å, respectively exhibit acceptable calibration plots, with the 1000 Å thick film having the closest slope value to the expected Nernstian slope. The PACPISE, on the other hand, exhibited a different trend; thin films of 500 Å thickness had closer to Nernstian behavior than those of greater thickness, as shown in Table I. The optimum slope and correlation coefficients for the calibration curves for the PMTCPISE and PACPISE are 58.55 mV/decade, and 1.00; and 51.18 mV/decade, and 0.997, for films having 1000 Å and 1500 Å thickness, respectively.

The effect of concentration of iodide in the doping solution was investigated using electrodes having 1000 Å-thickness films of PMT and PA, respectively. All dopant concentration studies were conducted on films doped at +0.8 V for 45 seconds. As can also be observed in Tables I(A) and I(B), increasing the dopant concentration results in a decreasing value of the linear dynamic range for the PMTCPISE and PACPISE. This deviation can probably be attributed to the increased oxidation of the electrode surface with the increase in the iodide ion concentration.

For a given film thickness of 1000 Å and a dopant solution concentration of 0.1M KI, the effect of the doping period on the linear dynamic range and the slope of the calibration curve of the PMTCPISE and the PACPISE were also investigated. For this given set of conditions, the optimum doping period was found to be in the range of 30 to 60 s, for both types of films, as illustrated in Tables I(A) and (B). The data presented so far indicate that in order to obtain an iodide–selective-electrode from the conducting polymer films with a wide linear dynamic range and a near Nernstian slope, there must be an optimum amount of iodide/iodine incorporated into the film matrix (which should have, in turn an ideal thickness). However, the amount of iodide/iodine should be restricted to the minimum amount required to generate a a surface redox potential within the polymer film. Exceeding such level of doping proved to be destructive to the physical integrity of the film resulting in an inferior response.

Thus, as the film thicknesses are increased, the concentration of the dopant required for an optimum electrode performance also increased. The relationship between the film thickness, the doping period, and the Nernst slope of the PMTCPISE for the determination of iodide ion is represented in Figure 2. Optimum response is found to occur for 1000 Å-thick films doped for a period of

Table I(A).   Effect of Various Preconditioning Parameters on the Correlation Coefficients (R) and Slope (S) of Calibration Graphs for Different Concentration Ranges for (a) PMTCPISE.

Parameter	Value	Concentration Range (M)			
		$10^{-1} - 10^{-5}$		$10^{-1} - 10^{-4}$	
		R	S	R	S
Film Thickness (°A)	250	0.996	57.86	0.995	57.33
	500	0.999	57.43	0.997	56.78
	1000	1.000	58.54	0.999	56.89
	1500	0.999	54.43	0.999	57.45
	2000	0.999	48.57	1.000	56.31
Dopant Concentration (M)	10^{-3}	0.999	56.10	0.999	54.5
	10^{-2}	1.000	57.35	0.999	53.64
	10^{-1}	0.998	58.81	1.000	58.54
	10^{1}	1.000	48.55	0.955	42.54
Doping Period (s)	30	1.000	57.98	1.000	58.54
	60	0.999	58.67	0.999	58.71
	120	0.999	56.78	0.999	55.46
	240	1.000	45.62	0.989	40.57
Doping Potential (mV)	700	0.999	58.56	0.998	57.43
	800	1.000	60.20	1.000	58.54
	1000	1.000	59.21	0.999	64.54
	1200	0.992	51.01	0.984	43.57
	1500	0.981	44.32	0.944	30.75

Table I(B)   Effect of Various Preconditioning Parameters on the Correlation
Coefficients (R) and Slope (S) of Calibration Graphs for Different
Concentration Ranges for (b) PACPISE.

Parameter	Value	Concentration Range (M)			
		$10^{-1} - 10^{-5}$		$10^{-1} - 10^{-6}$	
		R	S	R	S
Film	250	1.000	52.14	0.998	48.86
Thickness	500	0.999	51.67	0.997	51.18
(°A)	1000	0.997	49.04	0.990	45.11
	1500	0.995	48.11	0.990	40.82
	2000	0.997	42.21	0.995	38.04
Dopant	10^{-3}	0.999	54.11	0.999	41.50
Concentration	10^{-2}	0.999	50.88	0.997	42.14
(M)	10^{-1}	1.000	55.51	0.991	45.11
	10^{1}	0.985	45.16	0.966	34.39
Doping	30	0.999	50.34	0.991	45.11
Period (s)	60	0.998	50.12	0.998	44.00
	120	1.000	40.98	0.982	33.86
	240	0.998	32.45	0.927	29.89
Doping	700	0.999	47.71	0.981	43.14
Potential	800	1.000	48.46	0.991	45.11
(mV)	1000	0.998	40.20	0.985	36.98
	1200	0.992	30.06	0.969	26.75
	1500				

30 seconds in the 0.1 M KI/0.1 M $NaNO_3$ at +0.8 V. The film was also "exposed" to iodine vapors for a given period of time. The exposure of the surface was made possible using a "modified" cap. The cap contained iodine adsorbed on activated alumina or silica gel. This cap was used for "storage" or when the electrode was not in use in order to retain the "activity" and the response of the film for extended periods of time. The amount of iodine used was 0.0500 g, which was thoroughly mixed with 1.2000 g of silica gel or activated alumina. A diaphragm that contains 10 peripheral and one central orifices of 1 mm diameter each was

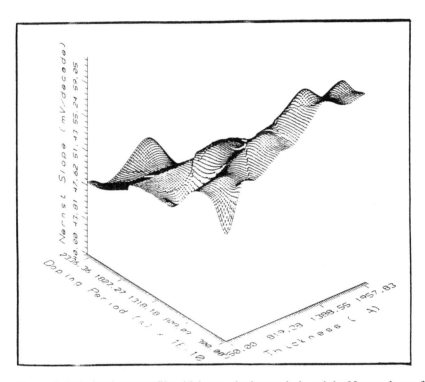

Figure 2 Relation between film thickness, doping period, and the Nernst slope of the PMT electrode for the determination of iodide ions.

used to separate the electrode tip from the iodine/alumina containing compartment. The cap-chamber capacity is 1.3 cc and was filled with the doping mixture just prior to the preconditioning step. The exposure of the PMT film to the iodine vapor for 30 s exhibited near Nernstian behavior and a larger linear dynamic range (1 x 10^{-7} to 1 x 10^{-1} M) than for films exposed for a period of time between 60 to 120 seconds. The previously observed damaging effect which results from prolonged periods of exposure to iodine under electrolysis conditions was also observed in the iodine vapor exposure study. Thus, the calibration curves obtained for films doped for periods >120 seconds showed a deviation from linearity in the calibration curves for concentrations ≤ 1 x 10^{-5} M iodide analyte and had lower slopes and correlation coefficients. Films preconditioned using a combination (hybrid) of the two methods showed superior characteristics over those preconditioned using only one method. A comparison of the three activation procedures are given in Table II. The most important advantages of the "hybrid" method of activation were observed in the lifetime of the electrode, reduced intereference by other ions, and improved (near Nernst) slopes for long exposure periods to the analyte solution. The results for the response time at the PMTCPISE and the PACPISE prepared by (a) the electrochemical doping, or (b) the hybrid electrochemical-chemical doping are given in Table II. For concentrations ≥ 1 x 10^{-5} M KI, the response time was in the range of 21 to 26 s with a standard deviation of ±2.45 for electrodes preconditioned using method (a), and 17 to 20 s with a standard deviation of ±0.67 for method (b), for a set of 14 electrodes each. A significant decrease of the response time of the CPISE preconditioned using method (b) was observed for iodide ion solutions with concentration $\leq 10^{-6}$ M. The electrodes preconditioned via method (a) showed a response time of ca. 47 to 54 s for the dilute solutions of iodide, and those prepared using method (b) displayed values of ca. 26 to 34 s.

The selectivity coefficients of the CPISE for the commonly encountered anions were determined by the commonly used method adopted by Rechnitz and Srinivasan [11], known by the "fixed interference method." The results for common interfering anions are summarized in Table II, which shows that for the large majority of anions the selectivity coefficients are of the order of 1 x 10^{-3} or smaller. It can also be expected that any redox system which can reduce iodine or oxidize iodide would interfere with the measurements.

The effect of temperature on the performance of the CPISE was examined for the electrodes preconditioned using the "hybrid" method. The temperature coefficients calculated from the data obtained are on the order of ca 0.09 to 0.15 mV/decade/K. Working temperatures > 80°C proved to be detrimental to the life time of the electrode. The drift of the potential to lower values can be interpreted as the leaching out of I_2 from the polymer body. This process apparently continues until a stable value for $[I_2]_f$ is reached or, alternatively it slows down. The electrode performance is unfortunately altered at that stage and the slope deviates from the expected Nernstian value. The electrodes preconditioned using the "hybrid" method, on the other hand, and kept in the cap containing iodine as the active agent displayed more stability and longevity.

Table II.   Effect of Preconditioning Parameters on the Characteristics of the PMTCPISE in the Determination of Iodide

*   In the hybrid method, both electrochemical and chemical doping were used. The electrode was kept at all times in a cap containing the activating agent.

Parameter	Value		
	Electrochemical	**Chemical**	**Hybrid***
Film Thickness (°A)	1000 R(0.999), S(56.78)	1000 R(1.000), S(57.31)	1000 R(1.000), S(58.31)
Dopant Concentration (M)	10^{-1} I^- R(1.000),S(58.54)	Sat. I_2 R(1.000),S(59.23)	Sat. I_2, 10^{-1} I^- R(1.000),S(58.75)
Doping Period (s)	60 R(0.999),S(58.71)	120 R(1.000),S(59.23)	60, 120 R(0.999),S(58.62)
Doping Potential (mV)	800 R(1.000),S(58.54)	---	800 R(0.999),S(58.62)
Lifetime (days)	15	23	>120
Selectivity Coefficient Cl^- Br^- NO_3^- ClO_4^- SCN^- $H_2PO_4^-$ SO_4^{2-} CO_3^{2-} $S_2O_3^{2-}$ OH^- CN^-	2.57 x 10^{-3} 4.71 x 10^{-3} 5.62 x 10^{-3} 2.12 x 10^{-3} 3.78 x 10^{-3} 4.61 x 10^{-4} 3.66 x 10^{-4} 2.79 x 10^{-3} 6.88 x 10^{-3} 2.30 x 10^{-1} 8.34 x 10^{-1}	3.55 x 10^{-3} 2.33 x 10^{-3} 1.66 x 10^{-4} 3.51 x 10^{-4} 5.67 x 10^{-4} 1.24 x 10^{-4} 2.71 x 10^{-4} 3.84 x 10^{-4} 2.33 x 10^{-3} 4.10 x 10^{-1} 6.45 x 10^{-1}	1.21 x 10^{-4} 4.56 x 10^{-4} 2.78 x 10^{-5} 5.16 x 10^{-5} 1.04 x 10^{-4} 4.66 x 10^{-5} 5.29 x 10^{-5} 1.34 x 10^{-4} 6.77 x 10^{-4} 1.18 x 10^{-1} 2.11 x 10^{-1}
Response Time (s)	21–26, σ(2.45)	20–25, σ(1.65)	17–20, σ(0.67)
Temperature Range (°C)	0–80	0–80	0–80

The selective determination of chloride, bromide and sulfite using a poly(-3-methylthiophene)–modified electrode was investigated in order to demonstrate the multifaceted nature of this class of conducting polymer as an ion–selective probe. The data obtained revealed that the chloride, bromide, and sulfite–selective–conducting polymer–based electrodes prepared by the above method have common features. These can be summarized as follows: (i) the linear dynamic range is limited (1×10^{-1} to 1×10^{-5} M concentration), (ii) the slope is in the range of 45 to 48 mV/decade, and (iii) the detection limits are in the order of 1.2-8.9×10^{-6} M. Moreover, their average lifetime was in the range of 2 to 3 days. The longevity of the electrode could only be extended upon keeping the electrode immersed in a 0.1 M solution of the ion under investigation. The "superior" performance of the conducting polymer film which was preconditioned for the potentiometric detection of iodide ion over those devised for other anions could be explained in terms of the nature of interaction of the film with the species under investigation. Thus, absorption of gaseous iodine by films of poly(thiophene) was found to be fully reversible, the doping process appeared to be similar to the case of conventional semiconductors, and I_2, the electrically active dopant, behaves as an acceptor [12]. The ion-selective electrode is affected by large hydrogen or hydroxyl ion activities, and the analyte under investigation may react with them. The iodide electrode, for example, responded over a limited pH range of 3.0 to 10.0 for 1×10^{-2} M I^- and 4.0 to 9.0 at the 1×10^{-4} M concentrations. The poor response beyond these limits could be explained as follows: at high pH values it responds to the hydroxyl ion (cf. the selectivity coefficient for hydroxyl ion in Table II.), while, at low pH, iodide (similar to fluoride) and hydrogen ions react to form undissociated hydrogen iodide.

Redox behavior at different polymer electrodes
Figure 3 shows the cyclic voltammetric behaviors of 5 mM catechol in 0.1 M sulfuric acid at PMT, PNMP, PAn and PF electrodes. All cyclic voltammograms displayed reversible redox peaks with distinctive peak separation values and different general potential–current features. From the analytical point of view, the sensitivity of the measurement could be regulated by the catalysis of the analyte/electrode charge transfer reaction. The comparison of the oxidation peak potential E_{ox} at these electrodes for 5 mM ascorbic acid in 0.1 M electrolyte (Table III) and of the oxidation for 5 mM catechol in 0.1 M electrolyte (Table IV) showed two important facts: (i) All polymer electrodes have relatively lower E_{ox} values as compared to those obtained at Pt or GC electrodes, (ii) for the series of polymer electrodes studied, the E_{ox} values increases in the order of PMT<PAn<PNMP<PF. The cyclic voltammograms (Figure 3) and the peak separations, ΔEp, of the anodic (E_{ox}) and cathodic (E_{red}) peak potentials (Table IV), indicates the extent of the reversibility of the redox process. The peak separation ΔEp displayed relatively smaller values for PAn when compared to PMT electrodes and noticeable larger difference with respect to the PNMP and PF electrodes. However, the cyclic voltammogram in Figure 3 for PAn electrode shows several peaks and represent complication in identifying the redox peaks for the analyte from those expected from the polymer film charge-discharge process. The PMT electrode, therefore, represents a relatively superior behavior over the other polymer films under consideration. The large current envelopes depicted in Figure 3 have been retained within all the polymer electrodes which is an inherent characteristics for this category of electrodes [13]. Moreover the

221

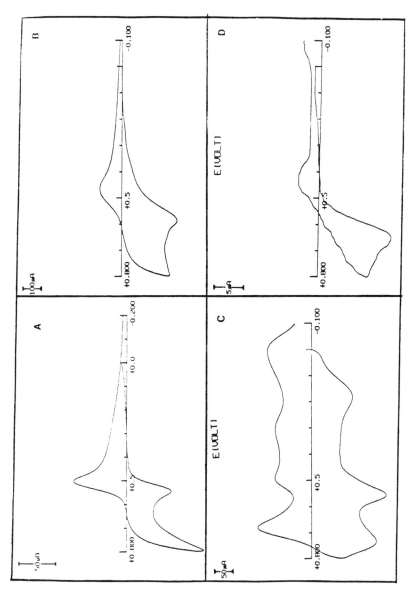

Figure 3 Cyclic voltammetry of 5 mM catechol in 0.1 M H_2SO_4, at P3MT (A), PNMP (B), PAn (C) and PF (D) electrodes. Scan rate 50 mV/s. All films prepared under similar conditions.

reversibility of the redox process was found to be superior for the PMT electrode as depicted by the amount of charge involved in the process, ca Q_{ox} = 28.04 mC/cm^2 and Q_{red} = 26.89 mC/cm^2. Careful inspection of Table IV reveals an interesting fact; the electrocatalytic activity of the polymer film is independent of the nature of the solid substrate used for the film deposition. Previous work [14] indicated that the selectivity of the detection was due to the permselective nature of the film and its porosity was mainly controlled by the synthesis conditions. Inspection of columns three and five of Table III indicate that the E_{ox} values for different supporting electrolytes studied are the same for films grown at GC or Pt surfaces. On the other hand, the oxidation potential values, E_{ox} are different for Pt and GC. This indicates that the oxidation process is taking place at the "surface" of the polymer film rather than at the substrate.

Table III: Electrolyte and Substrate Effect on the Electrochemical Response of PMT. Solution (5 mM ascorbic acid/100 mM Electrolyte). Scan rate 100 mV/s.

Electrolyte	pH	Pt	PMT/Pt	GC	PMT/GC
H_2SO_4	1.60	524(mV) 25 (μA)	445(mV) 45(μA)	663(mV) 67(μA)	439(mV) 170(μA)
Na_2SO_4	3.70	367 30	340 43	444 67	335 120
HCl	1.56	590 24	427 50	663 66	435 160
NaCl	3.23	439 11	381 22	512 42	388 130
HNO_3	1.25	568 23	422 36	667 62	430 160
$NaHO_3$	3.28	422 16	317 29	444 69	321 150
H_3PO_4	1.63	469 29	393 60	564 72	410 160
Na_3PO_4	3.85	385 19	291 37	475 46	298 130

Figure 4 shows the chromatogram obtained for a sample of a mixture of norepinephrine, L-DOPA, epinephrine and dopamine using PMT electrode. The conditions for the separation and detection were as follows: column; PARTISIL 5

ODS-3, mobile pbase; 0.15 M monochloroacetic acid, 50 mg SOS, pH 3.0, flow rate; 1.5 ml/min, electrochemical detector; and the PMeT electrode at an applied potential of 0.55 V (vs. Ag/AgCI). Equimolar amounts of the four analytes were used, namely 10-8 M. Comparison of this chromatogram to that shown in Figure S where a glassy carbon electrode was used reveals the following facts: (i) the current signals obtained in the case of using PMT as the working electrode is higher than those obtained in the case of using glassy carbon electrode, (ii) the base line of the first chromatogram shows a relatively higher stability when compared to that obtained in Figure 5. The use of PNMP as the electrochemical detector proved to dramatically affect the current signal and the base line stability.

Table IV. Electrochemical Data for Catechol at Different Conducting Polymer Electrodes. Catechol and Electrolyte concentrations were 5 and 100 mM, respectively. Scan Rate 100 mV/s. ΔEp is the Difference between Anodic and Cathodic Peal Potentials.

Electrolyte	PMT		PNMP		PF		PAn		Pt	
	Epa	ΔEp (mV)	Epa	ΔEp (mV)	Epa	ΔEp (mV)	Epa	ΔEp (mV)	Epa	ΔEp (mV)
H_2SO_4	560	69	587	122	643	183	556	67	702	364
Na_2SO_4	544	238	574	238	600	325	568	107	628	455
HCl	559	77	599	137	675	372	564	64	634	251
NaCl	590	250	618	285	689	464	598	218	610	376
HNO_3	559	79	578	121	673	323	55	67	658	278
$NaNO_3$	556	225	591	244	683	461	601	251	641	408
H_3PO_4	585	165	628	243	621	296	599	109	777	424
Na_3PO_4	513	231	604	211	605	385	581	291	616	558

Calibration curves for the catecholamines analyzed using the PMT and GC electrodes as the electrochemical detector are shown in figures 6a and 6b, respectively. The calibration curves describe the sensitive nature of the electrodes used towards the different catecholamines studied. The relative sensitivities and the detection limits (signal to noise ratio = 3) for the PMT and GC electrodes are given in Table IV. Again the PMT electrode showed relatively lower detection limits as compared to the GC electrode for the analysis of catecholamines. The effect of changing the flow rate on the current signals of the peak obtained in the chromatograms proved the relative fast response of the PMT electrode when

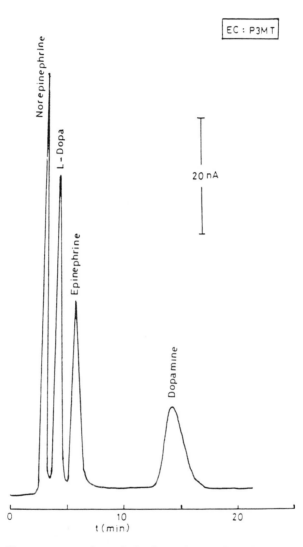

Figure 4 Chromatogram of catecholamines mixture. Sample volume 20 μL, PARTISIL 5 ODS-3 column, mobile phase 0.15 M monochloroacetic acid, pH - 3.00, 26°C, 100 nA full scale, flow rate 1.5 mL/min, chart speed 10 mm/min., detector BAS CV-1B, 0.55 V with P3MT working electrode.

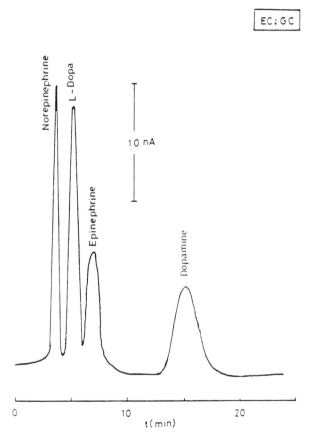

Figure 5 Chromatogram of catecholamines mixture. Conditions as in Figure 4 except GC used as working electrode, full scale is 50 nA.

226

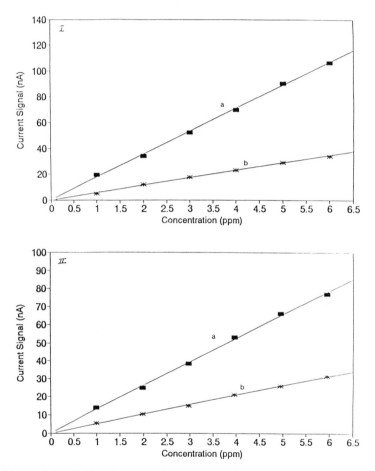

Figure 6 (a) Calibration curve for catecholamines mixture obtained under HPLC/EC condtions as indicated in Figure 4 at P3MT electrode for:
I. Norepinephrine; II. L-DOPA; III. Epinephrine; IV. Dopamine.
(b) Same as in (a) at a GC electrode.

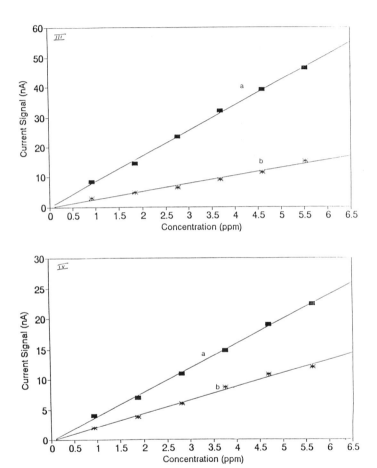

Figure 6. *Continued.*

compared to that of GC. The mechanism of response is similar when comparing the GC and PMT electrodes performances.

The PMT and GC electrodes were examined for the analysis of catecholamines over an extended period of time of over 60 days. The electrodes were also tested for a successive injection in the flow analysis mode for over 20 injections. The following observations can be noticed from the results obtained: (i) the current signal obtained at the PMT electrode is relatively more stable than that obtained at the GC, (ii) the current signal showed erratic behavior for successive injections, (iii) the current signal shows 35% attenuation (calculated as the ratio between 60th dayllst day current signal) for the PMT electrode compared to 69% attenuation for the GC. The results show a relatively high stability for the PMT electrode when compared to the GC electrode. The PMT electrode current signal showed only a relative decrease in magnitude of 0.7S% up to an extended time of use of over 60 days.

The stability of the electrode response was examined in the absence and presence of some surface active agents. Figure 7 shows the square wave voltammetry of 5 mM catechol in the presence of 1000 albumin at PMT, GC and Pt electrodes. The signal was attenuated in the case of Pt and GC electrodes and relatively unaffected in the case of using the PMT. The effect of repetitive cycles on the electrochemical response of GC and PMT was examined in the presence and absence of 1000 albumin for the analysis of p-aminophenol. The following observations were noticed: (i) the position of the oxidation peak potential was shifted to a more positive value upon the addition of albumin in the case of GC while relatively unaffected in the case of PMT, (ii) the effect of repetition in the absence of albumin attenuated the current signal after 100 cycles, on the other hand the PMT electrode showed only "partial" attenuation for the same number of repeated cycles. More interestingly the peak potential position relatively remained unchanged.

Table V. Relative Sensitivities and Limits of Detections of PMT and GC Electrode in the HPLC/EC Detection of Some Neurotransmitters. Analysis Conditions: 20 μL injection loop, 1.5 mL/min flow rate, 0.15 M (pH = 3.0) monochloroacctic acid mobile phase, Eapp=0.55 (vs. Ag/AgCI).

Analyte	PMT Electrode		GC Electrode	
	Sensitivities nA/mg/mL	LOD ng/mL	Sensitivities nA/mg/mL	LOD ng/mL
Norepinephrine	0.298	0.153	0.106	2.54
L–Dopa	0.757	0.137	0.349	1.46
Epinephrine	0.533	0.105	0.112	1.59
Dopamine	0.309	0.218	0.223	1.87

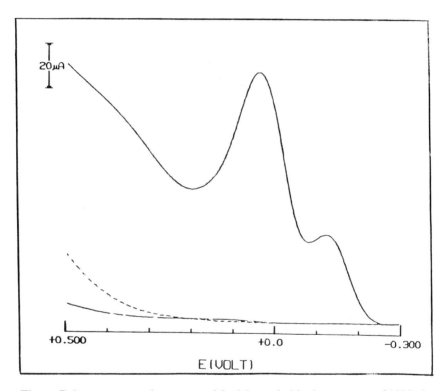

Figure 7 Square wave voltammetry of 5 mM catechol in the presence of 1000 albumin at PMT, GC, and Pt electrodes.

Conclusions:

The "hybrid" technique used for the "preconditioning" of the ion-selective electrode proved to extend the lifetime of the iodide selective electrode. The working temperature range, the selectivity coefficients, and response time of this sensor electrode are comparable to those commercially available. The preparation of bromide, chloride or sulfite selective electrodes was successful; however, extending lifetimes for such electrodes was difficult. The "unique" response of iodine-doped poly(3-methylthiophene) electrode towards iodide and its superior behavior over that of poly(aniline) or poly(N-methylpyrrole) is explained in terms of the specific interaction of iodine with the poly(3-alkylthiophene). The iodine-induced structural changes in the PMT may be different than that of other polymers. In this work we also compared the electrochemical behavior of some catecholamines at different conducting polymer electrodes. The electrodes were prepared under similar conditions; the results showed that the four polymers studied exhibited improved electrocatalytic activity when compared to GC or Pt electrodes. However, the PMT electrode showed better reversibility and stability over the other polymers studied. The detection limits obtained using the polymer electrode were as low as 0.1–.2 ng/mL. The polymer electrodes showed promising results for the resistance against surface fouling.

Literature Cited:

1. Murray, R.W., Ewing, A.G., and Durst, R.A., *Anal. Chem.*, **1987**, *59(5)*, 379A.
2. Kissinger, P. T., Hart, J. B. and Adams, R. N., *Brain Res.*, **1973**, *55*, 209.
3. Brazell, M. P., Kasser, R. J., Renner, K. J., Feng, J., Moghaddam, B. and Adams, R. N., *J. Neurosci. Meth.*, **1987**, *22*, 167.
4. Merz, A. and Bard, A. J., *J. Am. Chem. Soc.*, **1978**, *100*, 3222.
5. Michotte, Y., Moors, M., Deleu, D., Herrgobts, P. and Ebinger, G. ,*J. Pharm. Biomed. Anal.*, **1987**, *5*, 659.
6. Ishimitsu, T. and Hirose, S., *J. Chromatogr.*, **1985**, *337*, 239.
7. Tullk, K. and Pacakova, V., *Chem. Listy.*, **1979**, *68* ,800.
8. (a) Atta, N. F., Galal, A., Karagozler, A. E., Zimmer, H., Rubinson, J. F. and Mark, H. B. Jr., *J. Chem. Soc. Chem. Commun.*, **1990**, *19*, 1374, (b) Atta, N. F., Galal, A., Karagozler, A. E., Russell, G. C., Zimmer, H. and Mark, H. B., Jr., *Biosensors & Bioelectronics*, **1991**, *6*, 333.
9. Karagozler, A. E., Ataman, O. Y., Galal, A., Xue, Z., Zimmer, H. and Mark, H. B. Jr., *Anal. Chim. Act.*, **1991**, *248*, 163.
10. Galal, A., Atta, N. F., Rubinson, J. F., Zimmer, H. and Mark, H. B. Jr., *Anal. Lett.*, **1993**, *26(7)*, 1361.
11. Srinivasan, K. and Rechnitz, G. A. *Anal. Chem.*, **969**, *41*, 1203.
12. Kim, D. and Reiss, H., *J. Phys. Chem.*, **1985**, *89*, 2728.
13. Beck, F. and Oberst, M., *Macromol. Chem. Macromol. Symp.*, **1987**, *8*, 97.
14. Wang, J., Chen, S.-P. and Lin, M. S. *J. Electroanal. Chem.*, **1989**, *3*, 231.

Chapter 19

Fabrication and Evaluation of a Sulfide Microelectrode for Biofilm Studies

Tong Yu[1], Paul L. Bishop[1], Ahmed Galal[2], and Harry B. Mark, Jr.[3]

[1]Department of Civil and Environmental Engineering, University of Cincinnati, Cincinnati, OH 45221–0071
[2]Department of Chemistry, Faculty of Science, University of Cairo, Giza, Egypt
[3]Department of Chemistry, University of Cincinnati, Cincinnati, OH 45221–0172

In this chapter we describe the fabrication procedure and the evaluation of a solid-state ion-selective sulfide microelectrode for potentiometric measurement of total dissolved sulfide in aqueous solution at neutral pH. The sulfide microelectrode was made from a silver wire fused into a lead glass micropipette with a tip diameter of 20 μm. Its tip was chemically treated with $HgCl_2$ solution before being sulphidized with alkaline Na_2S solution. The sulfide microelectrode exhibited similar characteristics to that of the conventional sulfide electrode. The linear response of the microelectrode is in the range of 5×10^{-7} to 5×10^{-3} M sulfide in alkaline solution (pH=13.5). Its Nearnstian slope is - 31.1 mV per concentration decade. In neutral solution (pH 7.2) its linear response is in the range of 5×10^{-6} to 5×10^{-3} M total dissolved sulfide. Its slope in this range is -51.2 mV per concentration decade. The pH effect on the response slope of the sulfide microelectrode and the possibility of sulfate interference were also examined.

In this chapter we describe the fabrication and evaluation of a solid-state ion-selective microelectrode for the determination of total dissolved sulfide in aqueous solution at neutral pH. The total dissolved sulfide, S_T, is designated in this paper as the sum of dissolved hydrogen sulfide (H_2S), bisulfide (HS^-) and sulfide (S^{2-}). The need for such a microelectrode derives from the study of biological sulfate reduction in environmental biofilms, which typically have a neutral pH and a biofilm thickness about 1-3 mm. A microelectrode is needed to penetrate the biofilms to directly measure the concentration profiles of sulfide.

©1998 American Chemical Society

The commercially available sulfide electrode is too large for this application. It also requires the treatment of samples with sulfide anti-oxidant buffer (SAOB II), which raises the pH of the samples to above 12 (*1*). Since the early 1980's, researchers in microbial ecology have used sulfide microelectrodes to study microbial mats and biofilms (*2-5*). Their sulfide microelectrode was a solid-state Ag/Ag$_2$S ion-selective microelectrode. The tip diameter of the microelectrode was initially 200 μm, which was later reduced to 20-50 μm. It was used to measure biological samples at neutral pH. This microelectrode was made from a platinum wire fused into a lead glass capillary. The tip was ground flat and then recessed 30-50 μm by etching in a KCN solution. The recess was filled with silver by electroplating in a solution containing AgNO$_3$ and KCN. A layer of Ag$_2$S was formed on the surface of the silver by dipping the electrode into an (NH$_4$)$_2$S solution. However, the microelectrode was stable only for several hours due to the conversion of all the silver to Ag$_2$S (*2*).

In an effort to introduce and improve this microelectrode technique for environmental engineering biofilm research, we started our study of sulfide microelectrode with the basic structure of a modified Whalen's type oxygen microelectrode fabricated in our laboratory (*6-8*). The potential advantages of this microelectrode are its very small tip (approximately 5 μm diameter) and the ease of its fabrication. This microelectrode is made from a low melting-point (47°C) Bismuth alloy, consisting of Bismuth (44.7%), Lead (22.6%), Indium (19.1%), Tin (8.3%), and Cadmium (5.3%) (*9*). The tip of the microelectrode was chemically treated with or electroplated in an (NH$_4$)$_2$S solution under various conditions. However, the experimental results indicated that the potential responses of the electrode were not reproducible between measurements (*10, 11*). This lack of reproducibility may be caused by the complexity and uncertainty of the surface chemistry related with this alloy.

Exploring a new direction, we tried to fabricate the sulfide microeletrode using the procedure described in this paper, except for the last step. The pure silver tips of the microelectrodes, with and without a recess etched in a KCN solution, were chemically treated with 21.6% (NH$_4$)$_2$S solution for 2 min. Unfortunately, these microelectrodes did not respond very well to the concentration change of the sulfide standard solutions even at alkaline pH. Also, we noticed under the microscope that a fluffy black substance formed at the tip of the microelectrodes after repeated calibrations, possibly an indication of the dissolution of the silver (*11*). In the search for a better method to fabricate the sulfide microelectrode, we learned that Radić and Mark had observed an interesting phenomenon in their study of a sulfide millielectrode (*12*). They noted that, for a pure silver (99.99% Ag) wire, no black sulfide layer on the surface was observed even after six days of chemical treatment, and that only the silver wire containing 2.5-6.0% copper produced a silver sulfide layer after chemical treatment. They suggested that the silver sulfide layer is probably produced by an anodic reaction from a localized corrosion process. Dobcnik et. al. reported that the chemical pretreatment of pure silver wire with Hg^{2+} also produced a silver sulfide layer on the surface and a good sulfide ion-selective electrode (*13*).

In this study we: (a) developed a fabrication procedure for the sulfide microelectrode, (b) evaluated this procedure by calibrating the microelectrode at alkaline pH and by comparing it with a commercial sulfide electrode, (c) calibrated the sulfide microelectrode at the specified neutral pH range, and (d) analyzed the pH effect and examined possible interferences in its application. The sulfide microelectrode made using this procedure has a tip diameter of 20 μm. It is suitable for direct potentiometric measurement of sulfide concentrations in the range of 5×10^{-7} to 5×10^{-3} M in alkaline solution, and of total dissolved sulfide concentrations in the range of 5×10^{-6} to 5×10^{-3} M in neutral solution. It can be used for the measurement of sulfide in the study of biological sulfate reduction in environmental biofilms.

Fabrication

This sulfide microelectrode has a tip diameter of approximately 20 μm. It is made from a silver wire rather than a platinum one. The fine silver wire is etched in a KCN solution and fused into a pulled glass micropipette. The tip of the microelectrode is beveled, cleaned and chemically treated with first $HgCl_2$ solution and then Na_2S solution. The use of the silver wire makes the microelectrode body reusable after rebeveling the tip and repeating the chemical treatment as described in the last step of section **Fabrication**. In addition, the steps of making the recess at the tip by etching in a KCN solution and of filling the recess with silver by electroplating (*2*) can be eliminated. There are two key steps in the fabrication: first, to melt the glass such that it forms a good seal between the glass coating and the tip of the silver wire, and second, to pretreat the beveled tip with $HgCl_2$ solution before chemical treatment with the Na_2S solution. All of the chemicals used in this study were of analytical grade. The fabrication procedures for the sulfide microelectrode is as follows:

Pulling the Glass Micropipette. The lead glass micropipette (WPI, Catalog No. PG10150-4) was chosen because of its low softening point, good sealability and good insulating characteristics (*14*). The lead glass micropipette contains 22% PbO and has a softening point of 625°C. The commercially available micropipette has an o.d. of 1.5 mm, an i.d. of 0.75 mm, and a length of 10 cm. The center along the length of the micropipette was heated over the flame of a small burner until it became equally soft all the way around the cross-section at the heating point. This was done by constantly rolling the micropipette in the outer end of the flame between the fingers holding it. The micropipette was then taken quickly out of the flame and instantaneously pulled from both ends to about twice of its original length. As a result, the middle section of the micropipette became very fine but still unbroken.

Etching the Silver Wire. The original pure silver (99.99% Ag) wire has a diameter of 0.127 mm (Aldrich, Catalog No. 26,555-1). The tip of the wire was etched to approximately 10 μm in 2 M KCN solution (see section **Potassium Cyanide Solution** below). The setup is shown in Figure 1. The silver wire served as one electrode while the graphite rod (from the core of a No. 2 pencil) served as the other. A voltage between 2-4 V (a.c.) was applied. Most of the time, only the outermost 5-10 mm of the silver wire was immersed in the KCN solution and the wire was occasionally moved up and down to generate a smooth transition zone. The time at which the power is turned off is crucial to the final size of the tip. It seems that the silver wire was etched away section by section in the length of several millimeters. With good illumination, one can see with the eye exactly when the outermost tip disappears. The power should be turned off before the disappearance of the tip of the next section. The etched silver wire was then cleaned by immersing it sequentially in three beakers of Milli-Q water (18.2 cmΩ). The whole process of etching must be conducted in a highly ventilated hood. Refer to the Material Safety Data Sheet of KCN for safety precautions.

Potassium Cyanide Solution. 2M KCN solution was prepared by dissolving 6.512 g KCN in 50 ml Milli-Q water. A few particles of NaOH were added to the solution to keep it alkaline. Following the Material Safety Data Sheet of KCN for safety precautions.

Inserting the Etched Silver Wire. The etched silver wire was put on a microelectrode holder (Narishige, Catalog No. H-1) attached to a micromanipulator (World Precision Instruments Catalog No. M3301). The pulled glass micropipette was held horizontally by a clamp. The etched silver wire was very carefully inserted into the hollow center of the pulled glass micropipette at one end until the tip of the silver wire reached the very fine middle section of the micropipette. This process was closely watched through two magnifiers at right angles to each other, as shown in Figure 2. (There is only one magnifier showed in the figure.) It is essential that the fine tip is not bent during the insertion. Then the quality of the tip can be examined under the microscope.

Melting the Glass to Coat the Tip of the Silver Wire. The pulled glass micropipette with the etched silver wire inserted in it was hung vertically, with the etched tip of the silver wire pointing up. The glass micropipette was placed at the center of a trough heating filament (Sutter Instrument Co., Catalog No. FT330B), as shown in Figure 3. The heating filament was attached to a micromanipulator (World Precision Instruments Catalog No. M3301) and connected to an adjustable voltage transformer (Scientific Products, Catalog No. E2101-2). The heating filament was placed approximately several millimeters below the tip of the silver wire. A small rubber stopper was fitted around the lower end of the glass micropipette as a small weight. A small beaker with a piece of sponge covering the bottom was placed underneath the glass micropipette. The heating operation was monitored carefully through a horizontal microscope (World Precision Instruments, Catalog No. PZMH). Heat was applied slowly by turning the voltage transformer gradually up until the glass started to slowly melt. Heat was continuously increased as slowly as possible until the tip of the silver wire started to move downward. At this moment, the heat was suddenly increased. The lower half of the glass micropipette dropped into the beaker. The lead glass should form a thin coating on the tip of the silver wire and the tip should be sealed by the glass. It is crucial to slowly increase the voltage all the way through the heating operation. Turning the heat up suddenly at the last moment prevents the silver wire from breaking off before reaching the tip.

235

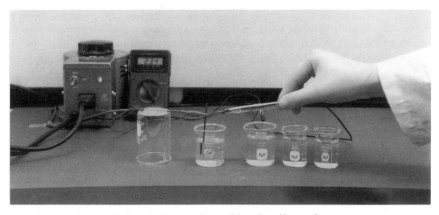

Figure 1. Set-up for etching the silver wire.

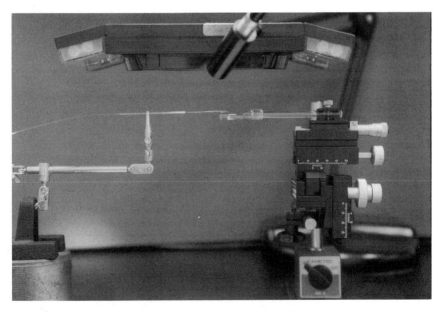

Figure 2. Set-up for inserting the etched silver wire into the pulled glass micropipette.

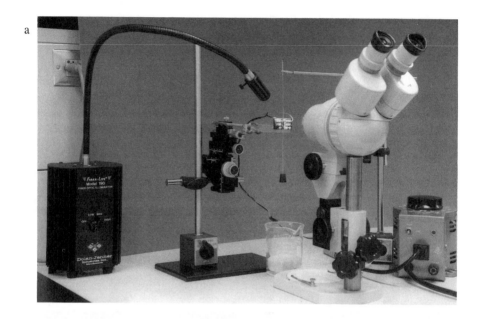

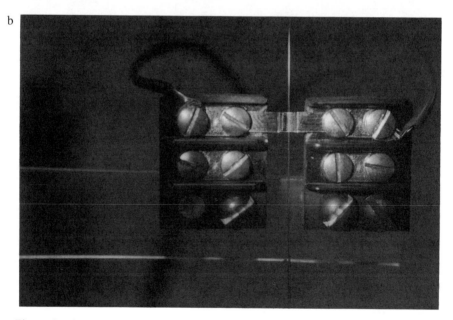

Figure 3. Apparatus and set-up for melting the glass to coat the tip of the silver wire. a) An overall view. b) Close-up: the heating filament and the tip.

Beveling the Tip. The prepared glass micropipette containing the silver wire was placed on a micropipette beveler (Sutter Instrument Co., Catalog No. BV-10). The tip was beveled on the diamond abrasive plate with a 45° angle to expose the silver surface at the tip. The beveled tip was examined under a microscope. A good seal between the glass coating and the tip of the silver wire is essential. An alternative method is to examine under the microscope if the silver surface at the tip is intact after the cleaning with HNO_3, as described in the last step of section **Fabrication**. A recess of the silver at the tip after the quick HNO_3 treatment usually indicates poor sealing (therefore, a much larger area for the silver to react with and to be corroded by the HNO_3).

Assembling the Microelectrode. A small piece of the low melting point Bismuth alloy (8) and then a fine electric wire were inserted into the glass micropipette from the tail end. The alloy was heated and melted to connect the silver wire and the electric wire. The electric wire was then connected to a BNC cable. The tail end of the glass micropipette was sealed with epoxy resin (Elmer's). The connection to the BNC cable was sealed with electric connector sealant (Archer, Catalog No. 278-1645).

Chemically Treating the Tip. The purpose of chemical treatment of the tip is to form an ion-selective sensing layer. The chemical treatment includes three steps: (1) Clean the tip of the microelectrode in an ultrasonic water bath (Fisher Scientific, Catalog No. 15-336) for approximately 20 seconds. Then treat it with 1:1 HNO_3. Finally, rinse it thoroughly with Milli-Q water. (2) Pretreat the tip with 0.1 M $HgCl_2$ (see section **Mercury Chloride Solution**) for 15 minutes. Then rinse it thoroughly with Milli-Q water. (3) Treat the tip with 0.1 M alkaline solution of Na_2S (see section **Sodium Sulfide Solution**) for another 15 minutes. Then rinse it thoroughly with Milli-Q water. Here, the chemical pretreatment of Hg^{2+} is essential to form the sulfide sensing layer. The less volatile Na_2S, instead of the $(NH_4)_2S$, was used to sulphidize the silver surface. Figure 4 shows the Scanning Electron Microscopic (SEM) pictures of the tip of the sulfide microelectrode after the above treatment. The raised center is the sensing layer on top of the silver surface. Surrounding it is the glass coating. The tip diameter of the sulfide microelectrode is approximately 20 μm. After the above treatment, the sulfide microelectrode was kept in water before use.

     Mercury Chloride Solution. 0.1 M $HgCl_2$ solution was prepared by dissolving 2.715 g $HgCl_2$ in 100 ml Milli-Q water.

     Sodium Sulfide Solution. The 0.1 M Na_2S solution was prepared as described in Sulfide Standard Solutions at Alkaline pH under Evaluation.

Evaluation

In order to evaluate whether the above fabrication procedure produces a good sulfide microelectrode, the sulfide microelectrode was first calibrated against sulfide standards at alkaline pH (as required by the instruction manual of the commercial sulfide electrode (1)). The calibration curve was compared with that of the commercial sulfide electrode to see whether both exhibit similar electrode characteristics. Then

a

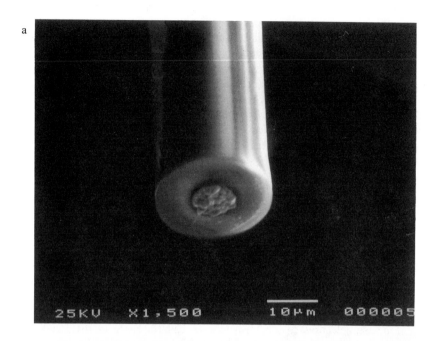

25KV   X1,500   10μm   000005

b

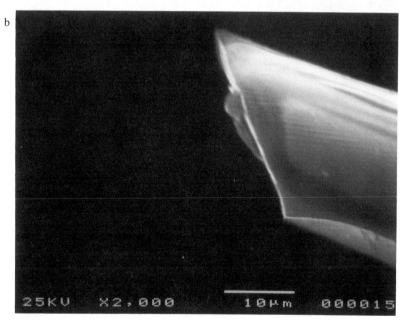

25KV   X2,000   10μm   000015

Figure 4. SEM pictures of the tip of the sulfide microelectrode. a) Front view: the raised center is the sensing layer formed on the surface of the silver and the surrounding is the glass coating. b) Side view: the raised center is the sensing layer formed on the surface of the silver and the tip is beveled with a 45° angle.

the microelectrode was examined for its electrode characteristics at neutral pH by calibrating against the total sulfide standards at the neutral pH. It was also compared with the commercial sulfide electrode calibrated at the same neutral pH. The result of the calibration at alkaline pH demonstrates that the sulfide microelectrode has similar characteristics to that of the commercial sulfide electrode and that the fabrication procedure used for making the sulfide microelectrode is valid. The result of the calibration at neutral pH shows that the sulfide microelectrode is suitable for direct potentiometric measurement of total dissolved sulfide at neutral pH. Finally, the effect of pH on the electrode characteristics and the possible interference under the application were also examined.

Calibration at Alkaline pH. The sulfide microelectrode was first calibrated at alkaline pH. The standard method D4685-92 of the American Society for Testing and Materials (ASTM) (*15*) and the Orion Instruction Manual for Sulfide Electrode Model 94-16 (*1*) were used as references. The sulfide microelectrode was calibrated against serial sulfide standard solutions at alkaline pH (see section **Sulfide Standard Solutions at Alkaline pH**). A Ag/AgCl millielectrode (Microelectrodes Inc., Catalog No. MI-409) was used as its reference electrode. The response of the sulfide microelectrode was also compared with that of the commercial sulfide electrode (Orion, Catalog No. 9416BN, tip diameter 12 mm) and its Ag/AgCl reference electrode (Orion, Catalog No. 900200). An Accumet Glass Body Combination pH Electrode (Fisher, Catalog No. 13-620-285) was used to monitor the pH in the standard solutions. These electrodes were connected to three Accumet Microprocessor Model 15 pH/mV meters (Fisher, Catalog No. 13-635-15A). These electrodes were immersed in each beaker of the sulfide standard solutions in turn. Between measurements, the electrodes were rinsed with Milli-Q water and the commercially purchased electrodes were blotted dry. It was found that the sulfide microelectrode started to deteriorate after overnight storage, regardless of whether it was stored in 0.1 M Na_2SO_4, Milli-Q water, or in air. Therefore, for each experiment in this study, the step of Chemically Treating the Tip in Fabrication, the preparation of the standard solutions, and the calibration of the sulfide microelectrode were all done on same day. We recommend that when measuring the concentration of sulfide in environmental biofilms the measurements should all be done on the same day as well.

Figure 5 shows the calibrations of the sulfide microelectrode and the commercial sulfide electrode at the alkaline pH. The upper curve is the calibration of the sulfide microelectrode. The filled dots are the mean of four sets of experimental data, while the range bars indicate the maximum and minimum values. The straight line is the least square linear regression of the four sets of data. The linear response range of the sulfide microelectrode is from 5×10^{-7} to 5×10^{-3} M, or from 0.016 to 160.3 mg/l as S. Its slope, -31.1 mV per concentration decade, is near Nernstian, compared to the theoretical value of -29.5 mV per concentration decade. The response time varied from 2 to 5 minutes. The pH values of the standard solutions were between 13.5 and 13.6 (not shown in the figure). The lower curve in the figure is the calibration of the commercial sulfide electrode. The filled squares are the mean of five sets of experimental data, while the range bars (barely seen) indicate the maximum and minimum values. The straight line is the least square linear regression of the five sets of data. The linear response range of the commercial sulfide electrode is from 1.125×10^{-7} to 1.125×10^{-2} M, or 0.0036 to 360.7 mg/l as S. Its slope is -28.3 mV per concentration decade. The response time (90%) was less than 1 minute. The pH values of the standard solutions were between 13.4 and 13.5 (not shown in the figure).

This calibration and comparison demonstrates that the sulfide microelectrode exhibits a similar slope and linear response range to that of the commercial sulfide electrode. The noticeable difference in the potentials between the sulfide microelectrode and the commercial sulfide electrode is likely due to the significant difference in the nature of the ion exchange matrix. The sensing layer, with a diameter less than 20 μm, at the tip of the sulfide microelectrode was generated in situ by letting the Ag wire chemically react first with $HgCl_2$ and then with Na_2S. The sensing layer generated in this way may have a relatively more porous structure and chemically different composition than that of the commercial sulfide electrode. The sensing layer of the commercial sulfide electrode was generated by first pressing Ag_2S crystals, under high pressure, to a pellet which has a diameter about 10 mm and then fixing it to the tip of the electrode. Therefore, the two electrodes have different sensing layers which indeed exhibit different electrode potentials.

Sulfide Standard Solutions at Alkaline pH. Precise sulfide standard solutions can not be prepared by weighing the sulfide salt because of the large and variable water of hydration (15). Therefore, the following procedure was used: (a) A stock solution of saturated sodium sulfide was prepared by dissolving approximately 100 g $Na_2S\cdot9H_2O$ in 100 ml deaerated Milli-Q water. This solution was shaken, securely stoppered, and allowed to stand overnight. A Na_2S solution, when acidified, can release hydrogen sulfide which is extremely toxic, even at low levels. Therefore, when dealing with Na_2S solution, it is important to follow the Material Safety Data Sheet of Na_2S for safety precautions. (b) A high concentration standard solution was prepared by first dissolving 16 g NaOH in 100 ml deaerated Milli-Q water in a 200 ml flask and allowing it to cool, then pipetting an exact amount (e.g. 10 ml) of the stock solution prepared in the previous step into the flask, and finally filling the flask to the mark using deaerated Milli-Q water. This standard solution contains approximately 0.1 M Na_2S and 2 M NaOH with a pH of about 13.5. It was then titrated with 0.1 M $Pb(ClO_4)_2$ (Orion, Catalog No. 948206), using the commercial sulfide electrode (Orion, Catalog No. 9416BN) as an indicator electrode, to determine its exact concentration of sulfide. The high concentration standard solution can be kept for a week (1). This 0.1 M standard solution of sulfide is also the 0.1 M Na_2S solution used to treat the tip of the sulfide microelectrode as described in section **Fabrication**. (c) The other standard solutions were prepared by serial dilution of the high concentration standard solution using equal volumes of 2 M NaOH and deaerated Milli-Q water. These standard solutions should be prepared with minimum aeration to avoid air oxidation of sulfide in the solutions. They should be prepared on the same day of the microelectrode calibration.

Calibration at Neutral pH. The pH has a significant effect on the chemical equilibrium of sulfide. Environmental biofilms are usually not at alkaline pH values, but rather at near neutral values. At neutral pH, the predominant species is no longer S^{2-} (16). But the total dissolved sulfide, S_T, designated in this paper as the sum of dissolved H_2S, HS^-, and S^{2-}, will still be the same amount as the S^{2-} measured at alkaline pH. The approach used to calibrate the sulfide microelectrode at neutral pH was to determine the S^{2-} concentration of the high concentration standard solution at alkaline pH and use this value as S_T at neutral pH. For biological sulfate reduction, the typical pH is 7.2 to 7.3 (5). Therefore, a phosphate buffer with pH 7.2 was used to prepare all standard solutions in this calibration. The sulfide microelectrode was calibrated against the serial total dissolved sulfide standard solutions at the neutral pH (see section **Total Dissolved Sulfide Standard Solutions at Neutral pH**). The experimental set-up for the calibration at neutral pH was the same as that at alkaline pH. The sulfide microelectrode and its reference millielectrode, the commercial sulfide electrode and its reference electrode, and the combination pH electrode were placed together and then immersed in turn in each beaker containing the standard solutions. Between measurements, these electrodes were rinsed with Milli-Q water and the commercially purchased electrodes were blotted dry.

Figure 6 is the calibrations of the sulfide microelectrode and the commercial sulfide electrode at the neutral pH. The upper curve is the calibration of the sulfide microelectrode. The filled dots are the mean of two repeated experimental measurements, while the range bars (barely seen) indicate the maximum and minimum values. The straight line is the least square linear regression of the two sets of data. The linear response range of the sulfide microelectrode at this pH is from 5.75×10^{-6} to 5.75×10^{-3} M, or from 0.18 to 184.3 mg/l as S. Its slope is -51.2 mV per concentration decade. Its response time (90%) was dependent on the total dissolved sulfide concentration and varied from 1 minute for the high concentrations to 2 to 5 minutes for the low concentrations. The pH values of all the standard solutions were at 7.2, except for the 5.75×10^{-3} M sample, where the pH changed to 7.4. (The pH data were not shown in the figure). The lower curve in the figure is the calibration of the commercial sulfide electrode. The filled squares are the mean of two repeated experimental measurements, while the range bars indicate the maximum and minimum values. The straight line is the least square linear regression of the two sets of data. The linear response range of the commercial sulfide electrode at this pH is also from 5.75×10^{-6} to 5.75×10^{-3} M, or from 0.18 to 184.3 mg/l as S. Its slope is -41.9 mV per concentration decade. Its response time is the same as that of the sulfide microelectrode at neutral pH. The pH values of the standard solutions were also the same as that of the sulfide microelectrode.

The result of this calibration and comparison shows that this sulfide microelectrode is suitable for measurement of total dissolved sulfide in aqueous solution at neutral pH. At the specified neutral pH, the sulfide microelectrode performed as well as the commercial sulfide electrode did. When the concentration of total dissolved sulfide was lower than 10^{-5} M, the sulfide microelectrode performed even better than the commercial sulfide electrode. At the specified neutral pH, the slope of the sulfide microelectrode, -51.2 mV per concentration decade, differs from the theoretical Nernstian slope, which is -29.5 and -59.0 mV per concentration decade for the dianion and monoanion reaction, respectively. These values indicate that the slope of the sulfide microelectrode is somewhere between the theoretical Nernstian slopes of the dianion and monoanion reactions, as would be expected as both anions are present at neutral pH.

242

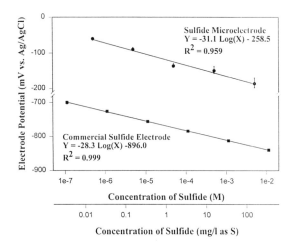

Figure 5. Calibrations of the Sulfide Microelectrode and the Commercial Sulfide Electrode at Alkaline pH. The upper curve is sulfide microelectrode and the lower curve is commercial sulfide electrode. Experimental data are expressed as • and ■, while ⎯⎯ is the corresponding regression.

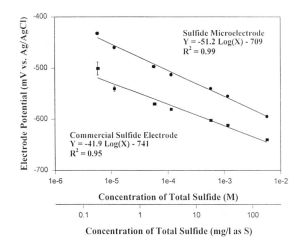

Figure 6. Calibrations of the Sulfide Microelectrode and the Commercial Sulfide Electrode at Neutral pH. The upper curve is sulfide microelectrode and the lower curve is commercial sulfide electrode. Experimental data are expressed as • and ■, while ⎯⎯ is corresponding regression.

Total Dissolved Sulfide Standard Solutions at Neutral pH. A phosphate buffer was used to maintain the pH of the standard solutions at 7.2. The buffer was prepared using NaH_2PO_4, Na_2HPO_4 and Mili-Q water (*17*) and was flushed with nitrogen gas continuously to deaerate it. The stock solution of saturated sodium sulfide and the high concentration standard solution at alkaline pH were prepared in the same way as described in (a) and (b) in section **Sulfide Standard Solutions at Alkaline pH.** Then, the high concentration standard solution at the neutral pH was prepared by first pipetting the exactly same amount of the stock solution as used for the high concentration standard solution at alkaline pH into a 200 ml flask, and then filling the flask to the mark using the deaerated pH 7.2 phosphate buffer. This high concentration standard solution contains the exactly same amount of total dissolved sulfide as that in the high concentration standard solution at alkaline pH. The other standard solutions were prepared immediately by serial dilution of the high concentration standard solution at the neutral pH using the deaerated pH 7.2 phosphate buffer. These standard solutions must be prepared just before the microelectrode calibration. The same caution should be applied to avoid air oxidation of sulfide in the solutions.

The Effect of pH. To examine the effect of pH, the sulfide microelectrode was calibrated at three different pH values in the neutral range: 7.0, 7.2, and 7.4. These values were chosen because they are centered at the specified neutral pH (7.2) for this study and they can all be maintained by phosphate buffer (*17*). The method of calibration was the same as that used in section **Calibration at Neutral pH.** All calibrations at the three pH values were conducted on the same day. Figure 7 shows the effect of pH on the calibration of the sulfide microelectrode. The upper, middle and lower curves are the calibrations of the sulfide microelectrode at pH 7.0, 7.2 and 7.4, respectively. The filled squares, dots and diamonds are the means of two repeated experimental measurements at pH 7.0, 7.2 and 7.4, respectively. The range bars (barely seen for most points) indicate the maximum and minimum values. The straight lines are the least square linear regression of the two sets of data for each pH value respectively. Since the calibration at pH 7.4 started at 1.15×10^{-5} M, for purposes of comparison the three curves show only the linear response range from 1.15×10^{-5} to 5.75×10^{-3} M. (The slope of the calibration at pH 7.2 was therefore slightly different from that in section **Calibration at Neutral pH.**) The slopes of this linear response range of the sulfide microelectrode are -55.4 mV per concentration decade for pH 7.0, -48.4 mV per concentration decade for pH 7.2, and -45.4 mV per concentration decade for pH 7.4. Table I summarizes the slopes of the sulfide microelectrode at four different pH values. These experimental results reveal that the slope of the sulfide microelectrode decreased from -31.1 to -55.4 mV per concentration decade while pH of the standard solutions decreased from 13.5 to 7.0. These intermediate slope values are exactly what would be predicted, assuming the use of normally accepted pK_a values for H_2S and HS^- ($pK_{a,1}$=7.02 and $pK_{a,2}$=13.9 (*18*)). For accurate measurement of total dissolved sulfide in biofilm samples, the exact pH value of the sample is a necessity.

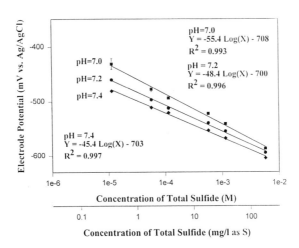

Figure 7. Effect of pH on the Calibration of Sulfide Microelectrode. The ■ and the upper curve are for the calibration at pH 7.0. The • and the middle curve are for the calibration at pH 7.2. The ◆ and the lower curve are the calibration at pH 7.4. Experimental data are expressed as ■, • and ◆ while ———— is corresponding regression.

Table I. Slopes of the Sulfide Microelectrode at different pH Values

	Slope of Linear Response Range (mV per concentration decade)
Dianion Nearnstian reaction	-29.5
pH = 13.5	-31.1
pH = 7.4	-45.4
pH = 7.2	-51.2
pH = 7.0	-55.4
Monoanion Nearnstian reaction	-59.0

Interference Study of Sulfate. Although most reported interfering species for sulfide electrodes (*1, 12, 13*) are not present at a significant level in biological sulfate reduction, sulfate, in equilibrium with sulfide, is abundant in the system. In order to examine whether sulfate interferes with the measurement of sulfide, the fixed interference method (*19*) was used. The sulfide microelectrode was calibrated in serial standard solutions at pH 7.2 with a fixed concentration of Na_2SO_4. The Na_2SO_4, after being dried at 103°C overnight, was added to the pH 7.2 phosphate buffer and its concentration was fixed at 5×10^{-3} M. The method of calibration was the same as that used in section **Calibration at Neutral pH**, except the sequence of measurement was from low concentration to high concentration standard solutions and then, without rinsing, repeated immediately from high concentration to low concentration standard solutions. The purpose is to examine whether there is concentration hysteresis on the sulfide microelectrode. The experimental results were compared with that of the calibration at pH 7.2 without Na_2SO_4. Figure 8 shows the calibration of the sulfide microelectrode at pH 7.2 with and without Na_2SO_4. The filled dots are the means of the two repeated experimental measurements at pH 7.2 with 5×10^{-3} M of Na_2SO_4, while the filled squares are the experimental measurements reported in section **Calibration at Neutral pH** (pH 7.2 without Na_2SO_4). The upper straight line is the least square linear regression of the calibration without Na_2SO_4 (the filled squares). Its slope is -51.2 mV per concentration decade. The lower straight line is the least square linear regression of the calibration with 5×10^{-3} M of Na_2SO_4 from 1.04×10^{-5} to 5.20×10^{-3} M of total dissolved sulfide. Its slope is -50.8 mV per concentration decade. The two lines are very close from approximately 1×10^{-5} to 5×10^{-3} M total dissolved sulfide. Considering that the two calibrations used two different sulfide microelectrodes and were conducted on two different days, the difference is within the

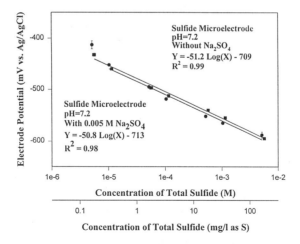

Figure 8. Calibration of the Sulfide Microelectrode at pH 7.2 with and without Na_2SO_4. The ■ and the upper straight line are the experimental data and the regression for the calibration without Na_2SO_4. The • and the lower straight line are the experimental data and the regression for the calibration with Na_2SO_4.

experimental error. Therefore, we conclude that sulfate does not interfere with the sulfide measurement using the sulfide microelectrode under these experimental conditions. This conclusion is also agree with the result of Dobcnik et. al. on their study of a sulfide millielectrode (*13*). As showed by the range bars of the filled dots in Figure 8, no concentration hysteresis was noticed with the sulfide microelectrode from 1.04×10^{-5} to 5.20×10^{-3} M of total sulfide.

Conclusions

The fabrication procedure described in this chapter produces a practical solid-state ion-selective sulfide microelectrode. The microelectrode has a tip diameter of 20 μm. It was made from a silver wire melted into a lead glass micropipette. Its tip was beveled and then chemically treated with $HgCl_2$ solution before being sulphidized with Na_2S solution to form a sensing layer.

The sulfide microelectrode can be used for potentiometric measurement of sulfide. At alkaline pH (pH >13.5), the sulfide microelectrode exhibits similar characteristics to that of the commercial Orion sulfide electrode. The linear response range of the sulfide microelectrode is from 5×10^{-7} to 5×10^{-3} M sulfide, or from 0.016 to 160.3 mg/l as S. Its Nearnstian slope is - 31.1 mV per concentration decade.

The sulfide microelectrode is also suitable for measuring total dissolved sulfide (designated as the sum of H_2S, HS^- and S^{2-}) in aqueous solution at neutral pH. At pH 7.2, the linear response range of the sulfide microelectrode is from 5.75×10^{-6} to 5.75×10^{-3} M total dissolved sulfide, or from 0.18 to 184.3 mg/l as S. Its slope in this range is -51.2 mV per concentration decade.

The pH has a significant effect on the characteristics of the sulfide microelectrode. The result of this study reveals that when pH decreased from 13.5 to 7.0 the microelectrode slope of the linear response range decreased from -31.1 to -55.4 mV per concentration decade. At neutral pH, the slope is typically in between the Nearnstian slope of a dianion and a monoanion reaction (29.5 and 59.0 mV per concentration decade, respectively). For accurate measurement of total dissolved sulfide at neutral pH, it is necessary to know the exact pH value.

Sulfate, in equilibrium with sulfide in biological sulfate reduction, does not interfere with the sulfide measurement.

Acknowledgment

This study was supported by the National Institute of Environmental Health Sciences, NIH and EPA (Grant number P42 ES 04908). Its contents are solely the responsibility of the authors and do not necessarily represent the official views of the NIEHS, NIH and EPA. The authors thanks Dr. William R. Heineman for valuble discussions.

References

1. Orion Model 94-16 Silver/Sulfide Electrode Instruction Manual.
2. Revsbech, N. P.; Jørgensen, B. B.; Blackburn, T. H. *Limnol. Oceanogr.* **1983**, *28*, 1062-1074.
3. Revsbech, N. P.; Jørgensen, B. B. In *Advances in Microbial Ecology*; Marshall, K. C., Ed.; Plenum Press: New York, London, 1986; Vol. 9, pp. 293-352.
4. Nelson, D. C.; Jørgensen, B. B.; Revsbech, N. P. *Appl. Environ. Microbiol.* **1986**, *52*, 225-233.
5. Kühl, M.; Jørgensen, B. B. *Appl. Environ. Microbiol.* **1992**, *58*, 1164-1174.
6. Whalen, W. J.; Riley, J.; Nair, P. *J. Appl. Physiol.* **1967**, *23*, 798-801.
7. Bungay, H. R., 3rd; Whalen, W. J.; Sanders, W. M. *Biotech. Bioeng.* **1969**, *11*, 765-772.
8. Fu, Y.-C. Ph.D. Dissertation, University of Cincinnati, Cincinnati, Ohio, 1993.
9. Belmont Alloy 2451 In *Alloy Digest*; Engineering Alloy Digest, Inc.: Upper Montclair, New Jersey, 1978; p. Bi-3.
10. Galal, A.; Yu, T.; Bishop, P. L.; Mark, J., Harry B. Presented at Pittcon'96, Chicago, Illinois, March 3-8, 1996.
11. Yu, T. Unpublished Experimental Data.
12. Radic, N.; Mark, Harry B., Jr. *Croatica Chemica Acta* **1985**, *58*, 179-188.
13. Dobcnik, D.; Gomiscek, S.; Stergulec, J. *Fresenius J. Anal. Chem.* **1990**, *337*, 369-371.
14. World Precision Instruments, 1995 Catalog, p. 55.
15. ASTM D4658-92.
16. Snoeyink, V. L.; Jenkins, D. *Water Chemistry*; John Wiley & Sons: New York, 1980; Chapter 4, pp. 86-196.
17. Gerhardt, P.; Murray, R. G. E.; Wood, W. A.; Krieg, N. R., Eds. *Methods for General and Molecular Bacteriology*; American Society for Microbiology: Washington, D.C., 1994; Chapter II, pp. 135-294.
18. Martell, A. E.; Smith, R. M. *Critical Stability Constants*; Plenum Press: New York, 1974.
19. Kluka. *Selectophore: Ionophores, Membranes, Mini-ISE*; Kluka Chemical Corp.: Switzerland, 1996, p. 11.

Chapter 20

Strategies for the Design of Biomimetic Oxoanion Ionophores for Ion-Selective Electrodes

J. Christopher Ball, Richard S. Hutchins[1], César Raposo[2], Joaquín R. Morán[2], Mateo Alajarín[3], Pedro Molina[3], and Leonidas G. Bachas[4]

Department of Chemistry, University of Kentucky, Lexington, KY 40506–0055

Biomimetic ionophores were developed for use in ion-selective electrodes (ISEs). By mimicking the strong interactions between certain biological molecules and oxoanions, ionophores have been designed that demonstrate high selectivity for particular oxoanions. Ionophores based on the guanidinium functional group and a derivatized urea group were prepared. ISEs for hydrogen sulfite, salicylate, and ibuprofen based on these ionophores are described.

Oxoanions (e.g., sulfate, sulfite, phosphate, and carboxylates) play an important role in several industrial processes, in living systems, and in the natural environment. Therefore, there is a need for simple analytical methods capable of selectively determining concentrations of oxoanions in various matrices. Ion-selective electrodes (ISEs) are well-suited in that regard. Polymer membrane ISEs are typically prepared by impregnating a plasticized polymer membrane with an ionophore. Ionophores (Greek for ion carriers) are synthetic receptors that interact selectively with particular ions. This interaction facilitates the partition of particular ions into the membrane, thus generating a potential that is directly related to the ion activity in the sample.

The selectivity of ISEs that are designed to respond to anions is typically evaluated by comparison to the Hofmeister series. The Hofmeister series is a ranking of anions based on their relative lipophilicity. ISEs that employ a non-selective ionophore, such as a quaternary ammonium salt, give a response pattern that follows the Hofmeister series: lipophilic organic anions > perchlorate > thiocyanate > salicylate > iodide > nitrate > bromide, nitrite > hydrogen sulfite > acetate, dihydrogen phosphate (1-3). Deviation of the response pattern of an ISE from the Hofmeister series requires a selective interaction of the anion with the ionophore.

---

[1]Current address: Pfizer Inc., Eastern Point Road, Groton, CT 06340.
[2]Current address: Departamento de Química Orgánica, Universidad de Salamanca, Salamanca, Spain.
[3]Current address: Departamento de Química Orgánica, Universidad de Murcia, Murcia, Spain.
[4]Corresponding author.

©1998 American Chemical Society

Ionophores can be designed to incorporate hydrogen-bonding sites to facilitate specific interactions with oxoanions. For example, synthetic receptors that incorporate secondary and tertiary amino groups with the proper spacing form strong complexes with oxoacids (the protonated forms of oxoanions) due to the presence of a bidentate hydrogen-bonding interaction. Such an interaction is shown in Figure 1 (Structure **1**) between 2-(acylamino)pyridine derivatives and carboxylic acids (4). The more acidic the hydrogens are made, the stronger the hydrogen-bonding. Protonating the pyridine in this type of receptor generates a positive charge, which can form a bidentate hydrogen-bonded ion pair with various oxoanions, including carboxylates (e.g., see Figure 1, Structure **2**) (4).

It is known that certain biological molecules (e.g., proteins) have the ability to bind to oxoanions. By mimicking the structure of these biological molecules, ionophores that have affinities for particular oxoanions can be obtained. These biomimetic ionophores can be incorporated into polymeric membranes to yield various oxoanion-selective electrodes. Two types of biomimetic receptors are presented here that are based either on the guanidinium functional unit or on a derivatized urea.

Natural Guanidinium-Based Receptors for Oxoanions

Proteins are a major class of biological molecules, several of which can bind oxoanions. The reason for this binding to oxoanions lies within the structure of the amino acids that compose these proteins. Recently, Copley and Barton conducted research into the structural properties that give some proteins their oxoanion binding ability (5). They analyzed the X-ray structures of 38 phosphate-binding proteins and 36 sulfate-binding proteins to determine where the binding sites occurred in the proteins. These studies showed that non-helical binding sites in the proteins were more specific in binding to oxoanions than were the helical binding sites. The amino acid that occurs with the greatest frequency in all of these non-helical, specific binding sites is arginine.

The ability of the guanidinium unit found in arginine residues of proteins to hydrogen-bond to phosphate and carboxylate groups (6-8) is depicted nicely in the organization of the binding site of the phosphate binding protein (9). This protein is found in the periplasm of Gram-negative bacteria (9). The X-ray structure of the complex of this protein with phosphate demonstrates that the guanidinium functionality of arginine-135 interacts simultaneously with both the complexed phosphate anion and the carboxylate group of aspartic acid-137 (Figure 2). An additional feature that is evident from the crystal structure is that the hydrogen bonds between the guanidinium functionality of arginine-135 and the carboxylate of aspartate-137 form a cyclic structure, which directs two of the amine hydrogens of guanidinium toward the binding cleft of the protein for hydrogen-bonding to phosphate (9). Cyclic guanidinium motifs can also be found in other natural guanidinium-based molecules, such as ptilomycalin A (Figure 3), a molecule produced by some warm water marine sponges (10). Studies that compare the binding of acyclic and cyclic structures containing the guanidinium functional group showed that the more rigid structure of the cyclic systems gave enhanced binding with oxoanions (11, 12). Synthetic receptors that make use of the guanidinium functionality have been designed for binding particular oxoanions, and subsequently used in our laboratory to develop oxoanion-selective electrodes (13).

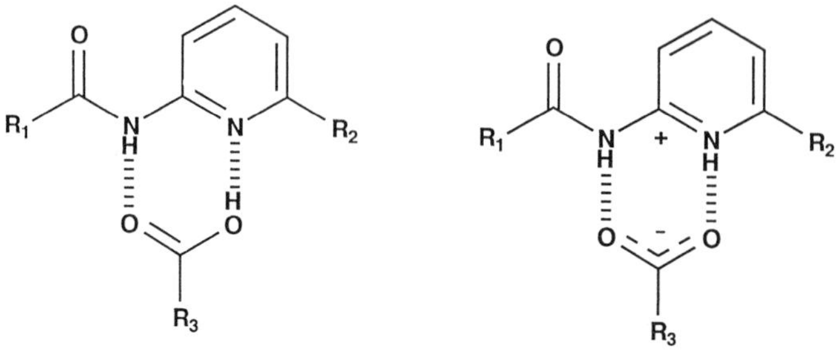

250

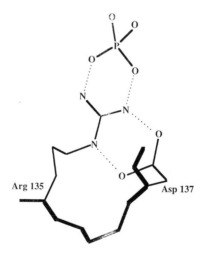

Figure 1. Hydrogen-bonding between 2-(acylamino)pyridine derivatives and carboxylic acids (**1**), and an example of a hydrogen-bonded ion pair consisting of a protonated 2-(acylamino)pyridine receptor and a carboxylate (**2**). (Adapted from reference 4)

Figure 2. Portion of X-ray structure of the complex between phosphate binding protein and phosphate. Thicker line denotes the protein backbone, while thinner lines denote the side chains of arginine-135 and aspartic acid-137. Hydrogens are omitted for clarity.

Urea as an Oxoanion Receptor

Urea and its derivatives have been shown to bind to oxoanions (14-19). Urea is derived directly from arginine in the last step of a biological process known as the urea cycle. The -NH groups in urea are responsible for its ability to hydrogen-bond to oxoanions in much the same manner as demonstrated with the guanidinium functionality. Urea as the molecule itself or as a unit of a larger receptor structure has been shown by X-ray crystallography and/or NMR titrations to bind carboxylates (15-17), sulfonates (17), and phosphates (17). The urea functionality has also been found in various bioaffinity systems. An example of this type of interaction is the binding of the ureido group of biotin to carboxylate groups of amino acid residues in streptavidin and biotin carboxylase (18,19).

In addition to the urea functional group, its sulfur analog, thiourea, shows binding to various oxoanions. In comparisons of these two functional groups, it has been demonstrated that the more acidic thiourea has a stronger binding towards oxoanions than urea (20, 21). In light of this, the replacement of the urea carbonyl by a stronger electron withdrawing group should yield a better hydrogen bonding receptor. This was demonstrated by Raposo et al. (22), who compared the oxoanion binding of a urea functional group fused to an aminochromenone (Figure 4, Structure **3**) with the binding of the same aminochromenone attached to a sulfuryl amide group (Figure 4, Structure **4**). In addition to the higher acidity of the sulfuryl amide hydrogen atoms, the geometry of the receptor is better suited for complexation of carboxylates. The sulfuryl-amide based receptor showed a higher association constant in binding to tetraethylammonium benzoate when compared to receptor **3** ($K_S = 20$ M^{-1} for **3** versus $K_S = 3.3 \times 10^2$ M^{-1} for **4**) (22). While the association constant of **3** is too low for the development of a useful sensor, the association constant of **4** is within the useful range for ISEs.

Results from Guanidinium-Based ISEs

Three guanidinium-based ionophores were prepared for use in ISEs (Figure 5) (for the synthetic details see reference 23). One of these guanidinium ionophores has a phenyl ring in a position that it is perpendicular to the hydrogen bonding site (Figure 5, Ionophore **5**). The ionophore was placed in a membrane with the composition of 1.0 mg of the BF$_4^-$ salt of **5**, 33 mg of poly(vinyl chloride) (PVC), and 66 mg of 2-nitrophenyloctyl ether (NPOE), a plasticizer. The studies were performed in 0.500 M Bis-Tris/HCl, pH 6.0. The ISE responded preferentially to HSO$_3^-$ and demonstrated a linear response in the logarithmic range of 5.0×10^{-5} to 0.50 M, with a slope of -47 ($\pm$2) mV/decade. The detection limit as defined by IUPAC (24) was found to be 3.9 ($\pm$ 0.3) $\times 10^{-5}$ M hydrogen sulfite. The data are averages $\pm$ 1 SD with n = 12. The ISE showed minimal interference from other anions (Figure 6). It is proposed that, in addition to the hydrogen-bonding interaction between the guanidinium group and the hydrogen sulfite oxoanion, there is an interaction between the phenyl group and sulfur, which contributes to the demonstrated selectivity for hydrogen sulfite over the other anions tested. Various parameters were studied to optimize and characterize the ISE response to HSO$_3^-$ (13). The electrode exhibited worsened detection limit when the pH of the test solution was increased from 6.0 to 8.0, but the slope of the calibration plot did not change significantly. Five different plasticizers were used to prepare membranes. All five gave preferential response to

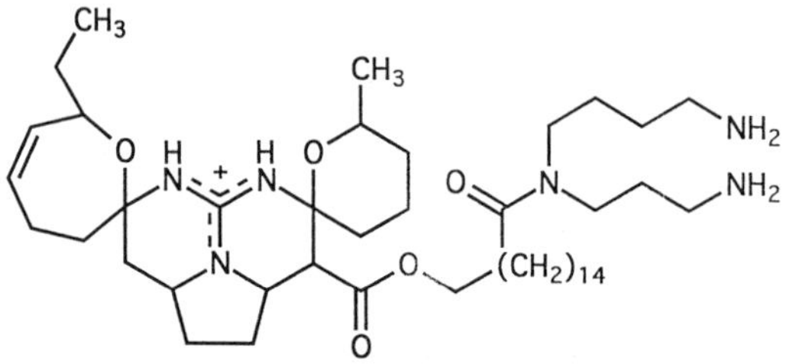

Figure 3. Structure of the natural product ptilomycalin A.

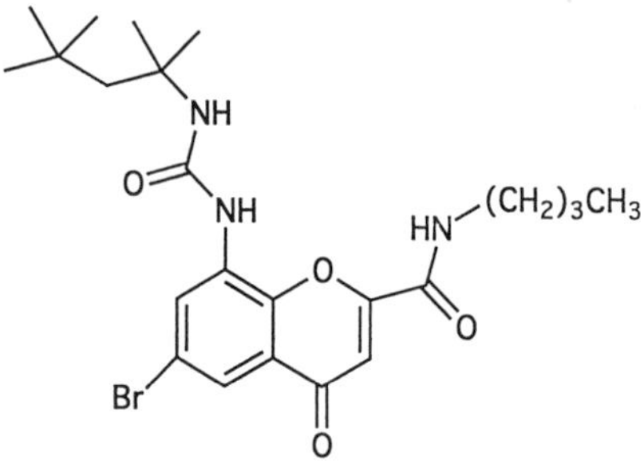

3

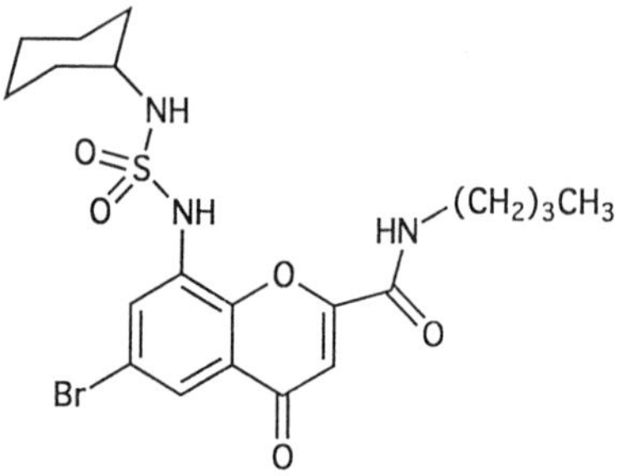

4

Figure 4. Structure of urea aminochromenone-based receptor (**3**) and sulfuryl amide aminochromenone-based receptor (**4**).

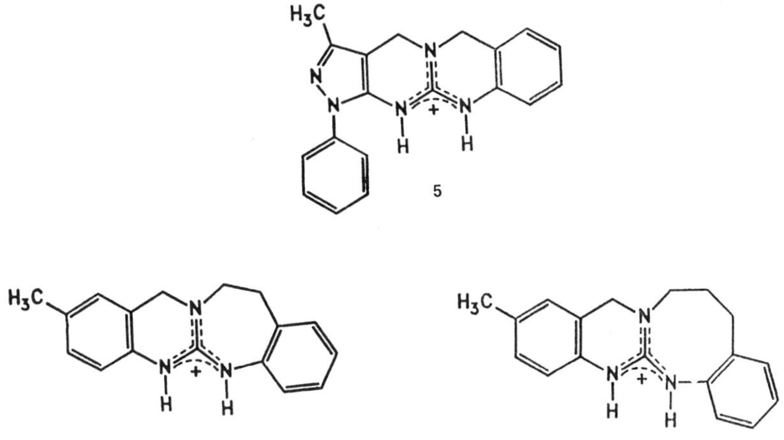

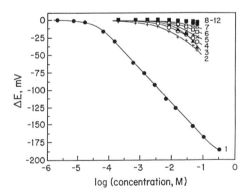

Figure 5. Structure of guanidinium-based ionophores.

Figure 6. Anion selectivity pattern of an ISE based on **5** • BF₄ in 0.500 M Bis-Tris/HCl, pH 6.0 buffer. Anions tested were hydrogen sulfite (1), perchlorate (2), salicylate (3), benzoate (4), thiocyanate (5), iodide (6), acetate (7), nitrite (8), thiosulfate (9), nitrate (10), sulfate (11), and phosphate (12). (Reproduced from reference 13)

HSO_3^-, with NPOE eliciting the strongest response. The lifetime, lipophilicity of the ionophore, and the binding mechanism were also studied (13).

The other two guanidinium-based ionophores studied (Figure 5, Ionophores **6** and **7**) lacked the attached phenyl ring. Both ionophores gave highest response to the salicylate anion. Salicylate is a carboxylate that is found in the human blood as a product of natural processes at concentrations not greater than 1.5×10^{-4} M. Acetylsalicylic acid is used as an analgesic, and the therapeutic levels of salicylate in serum range from 5×10^{-4} to 1.5×10^{-3} M (25). Ionophores **6** and **7** were used in membranes with the same composition as that given above for ionophore **5**, except that the ClO_4^- salts of the ionophores were used.

The response patterns of the ISEs prepared using **6** and **7** as ionophores differed greatly from the one obtained with ionophore **5**, and differed from each other with respect to the lesser responding anions. Both ionophores **6** and **7** gave ISEs with Nernstian responses for salicylate, the preferred anion. The ISEs with ionophore **6** gave calibration plots with slopes of -60 mV/decade, and the ISEs with ionophore **7** gave slopes of -61 mV/decade. The structural difference between ionophores **6** and **7** is that the latter has an additional methylene group. This makes **7** have a larger ring structure and be slightly more lipophilic. The lipophilicity value, log P, for ionophore **7** is 5.1, which is within the range of lipophilicities required for ISEs (26). The detection limit of ISEs prepared with ionophore **7** for salicylate (1.3×10^{-4} M) was an improvement over the detection limit obtained with **6** (6×10^{-4} M). The response patterns and selectivity of ISEs prepared using **6** and **7** are shown in Table I. In addition, high concentrations of chloride did not significantly affect the detection of salicylate.

Table I. Selectivity Coefficients of Salicylate-Selective Electrodes Prepared with Ionophores 6 and 7[a]

Interfering Anion	Ionophore 6	Ionophore 7
Benzoate	1.5×10^{-2}	3.0×10^{-2}
Acetate	7.1×10^{-3}	8.9×10^{-3}
Perchlorate	7.1×10^{-3}	5.2×10^{-3}
Nitrate	6.1×10^{-3}	4.7×10^{-3}
Thiocyanate	6.1×10^{-3}	5.3×10^{-3}
Hydrogen Sulfite	5.8×10^{-3}	4.7×10^{-3}
Phosphate	4.8×10^{-3}	4.2×10^{-3}
Iodide	4.0×10^{-3}	4.5×10^{-3}
Nitrite	3.7×10^{-3}	4.4×10^{-3}

[a] Selectivity coefficients calculated using the separate solution method at 4.0×10^{-2} M of anion.

Results from Sulfuryl Amide-Based Ibuprofen ISE

The sulfuryl amide-based receptor, **4**, was shown by NMR titration experiments to bind carboxylates (22). When this receptor was incorporated in a plasticized PVC membrane, it demonstrated selectivity for the ibuprofen anion. Ibuprofen is a well-known drug used for its analgesic and anti-inflammatory properties. The resulting ibuprofen ISE was evaluated to determine its detection limit and selectivity.

The detection limit of the ibuprofen ISE was found to be 1.3×10^{-4} M in a 0.0500 M MOPSO/NaOH buffer (pH 6.5). A Nernstian response was demonstrated by the electrode, giving a slope of -55 mV/decade of ibuprofen concentration. The selectivity coefficients for the ibuprofen ISE are shown in Table II. Further testing of the ISE is required to optimize the electrode for use in clinical applications.

Table II. Selectivity Coefficients of a Sulfuryl Amide-Based ISE for Ibuprofen

Interfering Anion	Selectivity Coefficients[a]
Iodide	3.2×10^{-1}
Thiocyanate	5.0×10^{-2}
Salicylate	3.2×10^{-2}
Perchlorate	$\leq 1.0 \times 10^{-2}$
Benzoate	$\leq 1.0 \times 10^{-2}$
Acetate	$\leq 1.0 \times 10^{-2}$
Chloride	$\leq 1.0 \times 10^{-2}$

[a] Selectivity coefficients calculated using the separate solution method II

Conclusion

The ability to design oxoanion ionophores based on biomimetic approaches for use in ion-selective electrodes has been realized. On the basis of the guanidinium recognition systems found in proteins and natural products, synthetic ionophores for hydrogen sulfite and salicylate have been successfully demonstrated using ionophores incorporating cyclic guanidinium functionalities. The recognition properties shown by urea and its derivatives have led to the development of a sulfuryl amide ionophore that is utilized in a membrane ISE for ibuprofen. By using natural binding phenomena as an inspiration, a variety of useful biomimetic ionophores may be realized for use in sensors with unique selectivity properties.

Acknowledgments

The authors would like to thank the following agencies for financial support of this work: the Kentucky Space Grant Consortium, the National Aeronautics and Space Administration, and the National Science Foundation.

Literature Cited

1. Hofmeister, F. *Arch. Exp. Pathol. Pharmakol.* **1888**, *24*, 247-260.
2. Chang, Q.; Park, S. B.; Kliza, D.; Cha, G. S.; Yim, H.; Meyerhoff, M. E. *Am. Biotech. Lab.* **1990**, *8*, 10-21.
3. Wotring, V. J.; Johnson, D. M.; Daunert, S.; Bachas, L. G. in *Immunochemical Assays and Biosensor Technology for the 1990's*; R. M. Nakamura, Y. Kasahara, and G. A. Rechnitz, Eds.; American Society of Microbiology: Washington, DC, 1992; pp 355-376.
4. Dixon, R. P.; Geib, S. J.; Hamilton, A. D. *J. Am. Chem. Soc.* **1992**, *114*, 365-366.
5. Copley, R. R.; Barton, G. J. *J. Mol. Biol.* **1994**, *242*, 321-329.

256

6. Price, P. A.; Williamson, M. K. *J. Biol. Chem.* **1985**, *260*, 14971-14975.
7. Riordan, R. F. *Mol. Cell. Biochem.* **1979**, *26*, 71-92.
8. Riordan, R. F.; McElvany, K. D.; Bordes, C. K. *Science* **1977**, *195*, 884-886.
9. Luecke, H.; Quiocho, F. A. *Nature* **1990**, *347*, 402-406.
10. Ohtani, I.; Kusumi, T.; Kakisawa, H.; Kashman, Y.; Hirsh, S. *J. Am. Chem. Soc.* **1992**, *114*, 8472-8479.
11. Dietrich, B.; Fyles, D. L.; Fyles, T. M.; Lehn, J. M. *Helv. Chim. Acta* **1979**, *62*, 2763-2787.
12. Dietrich, B.; Fyles, T. M.; Lehn, J. M.; Pease, L. G.; Fyles, D. L. *J. Chem. Soc., Chem. Commun.* **1978**, 934-936.
13. Hutchins, R. S.; Molina, P.; Alajarín, M.; Vidal, A.; Bachas, L. G. *Anal. Chem.* **1994**, *66*, 3188-3192.
14. Kelly, T. R.; Kim, M. H. *J. Am. Chem. Soc.* **1994**, *116*, 7072-7080.
15. Hamann, B. C.; Branda, N. R.; Rebek, J. *Tetrahedron Lett.* **1993**, *34*, 6837-6840.
16. Hughes, M. P.; Shang, M.; Smith, B. D. *J. Org. Chem.* **1996**, *61*, 4510-4511.
17. Smith, P. J.; Reddington, M. V.; Wilcox, C. S. *Tetrahedron Lett.* **1992**, *33*, 6085-6088.
18. Livnah, O.; Bayer, E. A.; Wilchek, M.; Sussman, J. L. *Science* **1989**, *243*, 85-88.
19. Waldrop, G. L.; Raymont, I.; Holden, H. M. *Biochemistry* **1994**, *33*, 10249-10256.
20. Fan, E.; VanArman, S. A.; Kincaid, S.; Hamilton, A. D. *J. Am. Chem. Soc.* **1993**, *115*, 369-370.
21. Nishizawa, S.; Bühlmann, P.; Iwao, M.; Umezawa, Y. *Tetrahedron Lett.* **1995**, *36*, 6483-6486.
22. Raposo, C.; Crego, M.; Mussons, M. L.; Caballero, M. C.; Morán, J. R. *Tetrahedron Lett.* **1994**, *35*, 3409-3410.
23. Molina, P.; Alajarín, M.; Vidal, A. *J. Org. Chem.* **1993**, *58*, 1687-1697.
24. Buck, R. P.; Lindner, E. *Pure Appl. Chem.* **1994**, *66*, 2527-2536.
25. *Tietz Textbook of Clinical Chemistry*, 2nd ed.; Burtis, C. A.; Ashwood, E. R., Ed.; W. B. Saunders: Philadelphia, PA, 1994; pp 1187-1191.
26. Dinten, O.; Spichiger, U. E.; Chaniotakis, N.; Gehrig, P.; Rusterholz, B.; Morf, W. E.; Simon, W. *Anal. Chem.* **1991**, *63*, 569-603.

Chapter 21

Using an 11-Ferrocenyl-1-undecanethiol Surface-Modified Electrode for Sensing Hydrogen-Ion Concentration in Concentrated Sulphuric Acid Solutions

Touma B. Issa[1], P. Singh[1], and M. Baker[2]

[1]School of Physical Sciences Engineering and Technology, Murdoch University, Murdoch 6150, Western Australia
[2]Department of Chemistry, University of Western Australia, Nedland 6907, Western Australia

The use of simple ferrocene/ferricenium couple in electrochemical sensors and other devices is limited because of its instability in aqueous media. However certain modified ferrocenes are found to be relatively stable and hence are attractive alternatives for the applications. This paper describes the electrochemistry of 11-ferrocenyl-1-undecanethiol ($FcC_{11}SH$) in sulphuric acid media, which is investigated by cyclic voltammetry of thin films of the material derivatised on gold substrates. It has been found that the oxidation and reduction peak potentials are independent of H^+ concentration in the pH range 2-10. However, the peak potentials change linearly with sulphuric acid concentration in the range 1-5 M. Extended cycling of the material indicated that, under the conditions of the experiment, $FcC_{11}SH$ was more stable than the simple ferrocene, and may be suitable for potentiometrically monitoring H^+ concentration in concentrated sulphuric acid solutions.

The determination of hydrogen ion concentration in dilute sulphuric acid electrolyte can be achieved by a pH measuring device such as the glass electrode. Unfortunately the glass electrode and other traditional pH sensors like antimony oxide electrode cannot be used in concentrated sulfuric acid solutions. Thus there is a need to develop alternative devices. We are exploring the possibility of using organic redox couples in developing a miniaturised device capable of monitoring the hydrogen ion concentration in concentrated sulphuric acid solutions.

Electrochemical studies on ferrocene (Fc) and its derivatives have been carried out extensively in the past few years (1). These compounds undergo reversible, one electron transfer reaction according to eqn. 1

$$Fc - e^- \rightleftharpoons Fc^+ \qquad (1)$$

©1998 American Chemical Society

The use of ferrocene/ferricenium couple as a reference electrode in non-aqueous media is well known (2,3). However, electrochemical potential sensing electrodes incorporating ferrocenes in aqueous media have to date suffered from a short working life due to susceptibility of the ferricenium ion (Fc^+) to hydrolysis and other decomposition reactions. The decomposition of ferricenium ions in aqueous media has been studied extensively (4-8). Although the results of these studies are obscure and often contradictory they generally agree that decomposition of the ferricenium ion occurs via displacement of the cyclopentadienyl ligands by nucleophilic groups from the reaction medium (6).

We have investigated the electrochemistry of thiol functionalised ferrocenes which are attached to the surface of gold, with a view to developing a ferrocene/ferricenium system which could be used in practical sensors and other devices. A specific aim of the study is to establish whether redox couples of these materials could be used as potentiometric sensors in aqueous solutions containing high concentrations (1-5M) of sulfuric acid .

This paper describes the results of our investigations on 11-ferrocenyl-1-undecanethiol ($FcC_{11}SH$).

Experimental

11-Ferrocenyl-1-undecanethiol ($FcC_{11}SH$), Figure 1, was synthesised in our laboratory according to the method reported in the literature (9-11).

For the electrochemical study, the electrodes (surface modified electrodes, SME) bearing a self-assembled monolayer (i.e., chemisorbed films) of the $FcC_{11}SH$ were prepared by dipping gold electrodes in dilute solutions of the ferrocene alkanethiol derivative in hexane (10). Surface modified electrodes consisting of gold electrodes covered with "physisorbed" ferrocene were prepared by the drop evaporation method (1). Prior to dipping, the gold surface was cleaned by polishing it with silicon carbide waterproof papers grades 800, 1200, then dipped in aqua regia for 1 minute. The cleaned electrode was then potentiostated for 5 minutes at -0.9 V vs SCE in 1.0M H_2SO_4 where hydrogen gas furiously evolved at the electrode surface. During these experiments the electrode was rotated at 1000 rpm to let the bubbles of the gas escape from the surface of the electrode. Before derivatising the gold electrode surface with the ferrocene of interest a background CV scan between the limits -0.1 V to 0.8 V vs SCE was run for each electrolyte. All the investigations were carried out under nitrogen atmosphere. The potentials were measured against a saturated calomel electrode (SCE) and quoted as such.

Results and discussion

Figure 2 shows a typical cyclic voltammogram (CV) of $FcC_{11}SH$ in 1.0M H_2SO_4. A well defined anodic and a corresponding cathodic peak consistent with a one electron reversible process were obtained. The surface coverage as calculated by integrating current under the oxidation portion of the CV was found to be 4.0×10^{14} molecule/cm^2. The theoretical monolayer surface coverage for the ferrocene, assuming the diameter of the ferrocene molecule to be 6.6Å, is calculated to be 2.7×10^{14} molecule/cm^2. Thus the experimental surface coverage was higher than a monolayer which could be related to some ferrocene molecules folded between the poorly packed polymethylene chains (12).

The difference in the corresponding anodic and cathodic peak potential was found to be ca. 30mV. This value should be zero for oxidation/reduction of a surface adsorbed species. However, Figure 3 shows that the plot of log anodic peak current Ip_a versus log scan rate is linear with a slope of 0.97 confirming that the electron transfer does involves surface adsorbed species (13). The observed peak separation possibly arises from the slow diffusion of the counter anion through the ferrocene layer at the electrode surface.

The pH dependence of Ep_a of $FcC_{11}SH$ is shown in Figure 4. The results for ferrocene are also shown for comparison. As expected, the Ep_a values for $FcC_{11}SH$ are independent of pH in the range 2-10.

The effect of higher concentrations of sulfuric acid (1-5M) on the anodic peak potential of the various ferrocenes is shown in Figure 5. As can be seen the anodic peak potentials shift linearly to less positive values as the acid concentration is increased. The variation of the peak potential of this ferrocene with concentration of H_2SO_4 is reported to be due to the liquid junction potential of the reference electrode (14). Although this is a logical explanation, it does not account for the fact that the slopes of Ep_a vs $[H_2SO_4]$ plots for ferrocene (50 mV) and $FcC_{11}SH$ (29 mV) are quite different. If liquid junction potential were the only mechanism the two slopes should have been identical. Some other ferrocenes which have been investigated under identical conditions in our laboratory also show a range of slopes of their Ep_a vs $[H_2SO_4]$ curves.

While further work is in progress to determine the cause of the observed potential shift with concentration of H_2SO_4, the important thing to note here is that whatever may be responsible for the observed results, the results could form the basis of developing a sensor for monitoring high concentration of H_2SO_4 such as those found in the lead-acid battery. Hence the measurement of state of charge of lead-acid batteries should be possible by use of a combined Fc/SCE electrode.

The stability of the surface modified electrodes was investigated as follows. The electrodes (SME's) of ferrocenes of interest were prepared as described in the experimental section. The SME's were then polarised anodically in 1.0M H_2SO_4 to achieve a partial oxidation of the material. The polarisation was switched off and the equilibrium potential of the SME was monitored against a saturated calomel electrode as a function of time. Typical results are shown in Figure 6. As can be seen the potential of both the ferrocenes drops rapidly at the start, while the oxidised and reduced forms approach equilibrium. The potentials soon reach a steady value. Once steady values had been obtained, the SME's were subjected to cyclic voltammetric scans at several intervals of the experiment.

Anodic and cathodic peaks similar to those shown in Figure 2 were reproduced for both the ferrocenes. However, while the anodic peak current of $FcC_{11}SH$ remained constant that of ferrocene dropped of rapidly. This is not unexpected, because Fc^+, being soluble, falls off the electrode surface. The effect of repeated cycling on the anodic peak currents (Figure 7) can also be explained in the

Figure 1. 11-ferrocenyl-1-undecanethiol $FcC_{11}SH$

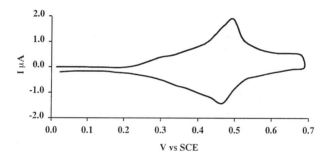

Figure 2. Cyclic voltammogram of $FcC_{11}SH$ in 1.0M H_2SO_4 at a gold electrode (sweep rate 100mV/sec).

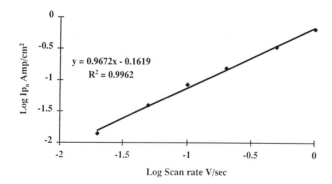

Figure 3. Log-Log plot of the peak current versus scan rate for the oxidation of $FcC_{11}SH$ with a slope of 0.9672.

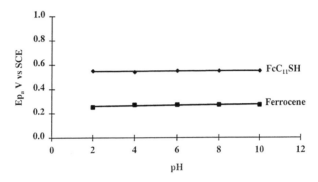

Figure 4. Epa versus pH for Ferrocene and $FcC_{11}SH$ in 0.1M K_2SO_4.

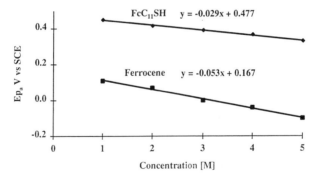

Figure 5. The anodic peak potential versus concentration of H_2SO_4 for Ferrocene and $FcC_{11}SH$.

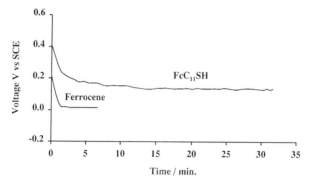

Figure 6. Equilibrium potential of Fc⁺/Fc couple in contact with 1.0M H_2SO_4 as a function of time for compound Ferrocene and $FcC_{11}SH$.

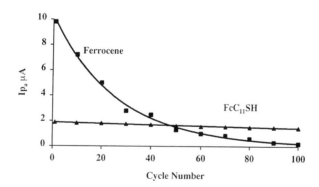

Figure 7. Anodic peak potential of compound $FcC_{11}SH$ and Ferrocene after 100 cycles.

same way; Fc drops off the electrode surface due to the solubility of Fc^+ in the medium, whereas $FcC_{11}SH$ is retained on the surface on repeated cycling.

Note that initially Ip_a of Fc is greater than that of $FcC_{11}SH$ because the surface coverage of the ferrocene is higher due to its preparation method.

Conclusion

The cyclovoltammetric study of the ferrocene derivative in acidic aqueous solutions shows that the $FcC_{11}SH$ undergoes a one electron redox reaction. The anodic and cathodic peak potentials are independent of the solution pH in the range 2-10. However in more concentrated acid solution the peak potentials vary linearly with concentration of the acid in the range 1-5 M. The variation is approximately 30 mV per unit molar concentration of the acid. The equilibrium potential of the investigated Fc^+/Fc couples coated on gold in contact with aqueous acidic solutions rapidly reached a constant value. On repeated cycling the peak current dropped off fairly quickly for Fc but remained constant for $FcC_{11}SH$ for at least 100 cycles. This suggests that the electroactive species of Fc are either soluble in the media or are unstable. However, $FcC_{11}SH$ is reasonably stable in concentrated acid solutions.

Literature Cited

1. Murray, R. W. In *Chemically Modified Electrodes*; Bard, A. J., Ed.; Electroanalytical Chemistry; Dekker: NewYork, **1984** Vol.13; pp 191-368, and references therein.
2. Peerce, P. J.; Bard, A. J. *J. Electroanal. Chem* **1980**, *108*, 121.
3. Gritzner, G.; Kuta, *J. Pure Appl. Chem.* **1984**, *56*, 461.
4. Matveev, V. D. ; Poroshenko, I. G. *J. Gen. Chem. USSR.* **1989**, *59*, 533.
5. Matveev, V. D. ; Poroshenko, I. G. *J. Gen. Chem. USSR.* **1989**, *59* , 41.
6. Holecek, J.; Handlir, K.; Klikorka, J.; Bang, N. D. *Coll. Czech. Chem. Commun.* **1979**, *44*, 1379, and references cited therein.
7. Aly, M. M. *Inorg. Nucl. Chem. Lett.* **1973**, *9*, 369.
8. Prins, R.; Korswagen, A. R.; Kortbeek, A. G. T. G. *J. Organomet. Chem.* **1972**, *39*, 335.
9. Hickman, J. J.; Ofer, D.; Laibinis, P. E.; Whitesides, G. M.; Wrighton, M. S. *Science* **1991**, *252*, 688.
10. Hickman, J. J.; Ofer, D.; Zou, C.; Wrighton, M. S.; Laibinis, P. E.; Whitesides, G.M. *J. Am. Chem. Soc.* **1991**, *113*, 1128.
11. Okahata, Y.; En-na, G.; Takenouchi, K. *J. Chem. Soc., Perkin II* **1989**, 835.
12. Chidsey, C. E. D.; Bertozzi, C. R.; Putvinski, T. M.; Mujsce, A. M. *J. Am. Chem. Soc.* **1990**, *112*, 4301.
13. Gomez, M. E.; Kaifer A. E. *J. Chem. Ed.* **1992**, *69*, 501.
14. Fischer, A. B. Thesis, Massachusetts Institute of Technology, Cambridge (**1981**).

FIBER OPTICS SENSORS

Chapter 22

Real Time pH Measurements in the Intact Rat Conceptus Using Ultramicrofiber-Optic Sensors

Weihong Tan[1,3] Bjorn A. Thorsrud[2,4], Craig Harris[2], and Raoul Kopelman[1,5]

[1]Department of Chemistry, University of Michigan,
Ann Arbor, MI 48109–1055
[2]Toxicology Program, Department of Environmental and Industrial Health,
University of Michigan, Ann Arbor, MI 48109–2029

Real time measurements of pH have been made in single, viable, intact rat conceptuses during the period of organogenesis, using newly developed ultramicrofiberoptic chemical sensors. The fiberoptic sensors are constructed using a novel technology based on photo-nanofabrication of optical probes and near-field photochemical synthesis. The pH sensors are a thousandfold smaller, with a millionfold reduction in necessary sample size, and show at least a hundredfold reduction in response time, when compared to other optical sensors currently available. In the present application, biosensors are inserted into the extraembryonic fluid (EEF) compartment of an intact rat conceptus, which is positioned in a customized perifusion chamber. Viability is maintained and exposures are controlled during pH determinations, while causing no damage to or leakage of fluid from the surrounding visceral yolk sac (VYS). Static determinations of pH in intact rat conceptuses of varying gestational ages showed decreasing pH with conceptal age. Dynamic alterations in pH were also measured in response to several variations in environmental conditions and specifically during exposure to a cellular thiol oxidant, diamide.

An understanding of the functional, dynamic responses of living cells and complex organisms to endogenous and exogenous signals has long been a goal of biologists from both fundamental and applied disciplines. Many traditional methods for determination of cellular, biochemical and molecular functions have, of necessity, been invasive, requiring destruction of the cell or organism. Micro-optical probes and ultra small microelectrodes have been developed to study minute functional biological units *in vivo* and *in vitro* (1-6), including noninvasive monitoring of biochemical events in periportal and pericentral hepatocytes in the intact liver (7,8). Generally, measurements by conventional methods have been determined exogenously to the

---

[3]Current address: Department of Chemistry, University of Florida, Gainesville, FL 32611.
[4]Current address: Department of Chemistry, University of Michigan, Ann Arbor, MI 48109–1055.
[5]Corresponding author.

©1998 American Chemical Society

biological sample(*4,5*) or following tissue disruption (*9,10*). One of the limitations of these conventional techniques has been the absence of probes small enough to be used in extremely small, fragile and dynamic biological systems, such as early embryos and single cells. Irrespective of these specific limitations, the use of fiberoptic chemical biosensors has been growing for both environmental and biological applications (*1,2*). However, fiberoptic sensors are usually larger than 100 µm and are thus inappropriate for single cell operations. In addition, response times have been relatively long, typically in the range of seconds to minutes, making determinations of rapid responses impractical. We have recently developed ultrasmall optical fiber sensors (*11,12*) for spatially resolved measurements in small organisms and single cells. Sensor sizes have been reduced down to 0.1 µm and have sample volume requirements of only femtoliters. Additionally, these sensors are capable of very rapid (millisecond) monitoring of chemical and biological reactions. Ultrasmall optical fiber sensors are prepared using a photo-nanofabrication technology based on near-field photochemical synthesis and nanofabricated optical probes (*11-13*), utilizing optical fiber tips (*13,14*). The working range of optical fiber sensors includes probes with tip dimensions in the submicrometer to tens of microns range. They are fabricated by incorporating a derivative of fluoresceinamine, N-fluoresceinylacrylamide (FLAC) into an acrylamide and N,N-methylenebis (acrylamide) copolymer that is attached covalently to a silanized fiber tip surface through near-field photopolymerization (*11*). The insert in Figure 1 shows a schematic drawing of the fiberoptic pH sensor tip. In order to enhance the working ability of the miniaturized fiber optic sensors, internal calibration methods have been developed (*12*) and are based on the fluorescence intensity ratios obtained from different wavelengths of the same emission spectrum for a single dye elicited by a single excitation signal. This procedure is highly effective for small-sized sensors, especially when dye species absorption differences are also utilized (*12*).

The first significant biological applications of the ultramicrofiberoptic chemical sensors are demonstrated here using the rat whole embryo culture system. Embryo explantation, growth and culture conditions are previously described (*10*). Explanted, cultured, viable, gestational day 10 to 12 (GD 10-12) rat conceptuses (*15,16*) are carefully placed inside a customized perifusion chamber, which is positioned on the stage of an inverted fluorescence microscope, as shown in Figure 1. This apparatus is able to maintain the viability of GD 10-12 rat conceptuses in serum-free medium for over an hour of monitoring. A schematic of how the miniaturized fiberoptic chemical sensor is inserted and positioned in the EEF space of the conceptus is shown in Figure 2. The spectroscopic protocol is described in detail (*12*). A calibration curve was obtained from pH measurements of Hank's Balanced Salt Solution (HBSS). Preliminary experiments, using HBSS, without the embryo present, verified the ability of the submicrometer fiberoptic sensor to discriminate 0.1 pH unit changes in the pH range of 6.6 to 8.6. Next, GD 10 and 12 conceptuses were analyzed for differences in pH as a function of advancing gestational age. EEF pH measurements, determined by using the fiber optic sensor, show values of 7.50 to 7.56 in the 10-16 somite, GD 10 rat conceptus and pH values of 7.24 to 7.27 in the 32-34 somite, GD 12 conceptus (see Table 1). These values are in good agreement with previous results obtained by using large numbers of homogenized mouse embryos at comparable stages of development, where pH was determined via a radiochemical accumulation method (*10*). Our new method, in contrast, uses a single, live conceptus for the pH measurements. The use of a single embryo has numerous advantages over pooled embryos, in that it is possible to maintain structural and functional integrity, while monitoring static and dynamic pH changes in the EEF. The conceptus also serves as its own control, can be accurately characterized as to developmental stage, can be monitored spatially and temporally in real time and can be returned to whole embryo culture to evaluate other relevant endpoints later in development.

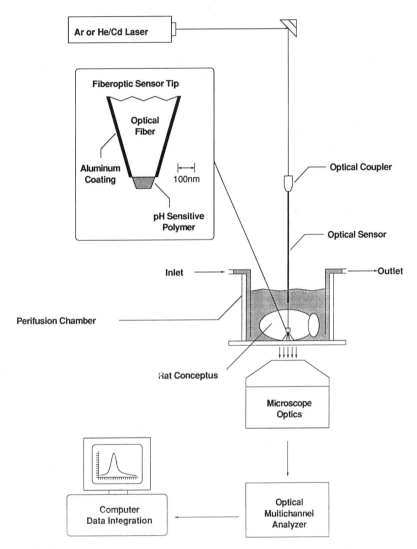

Figure 1. Schematic diagram of the fiberoptic perifusion system, depicting the intact rat conceptus in the customized monitoring chamber, located on the stage of an inverted fluorescence microscope. The emissions signal originating from the tip of the pH sensor inserted into the extraembryonic fluid compartment is detected through the microscope optics, directed to an optical multichannel analyzer (OMA) and integrated using a micro computer. The tip of the ultramicrofiberoptic pH sensor has been amplified and is shown schematically in the figure insert.

269

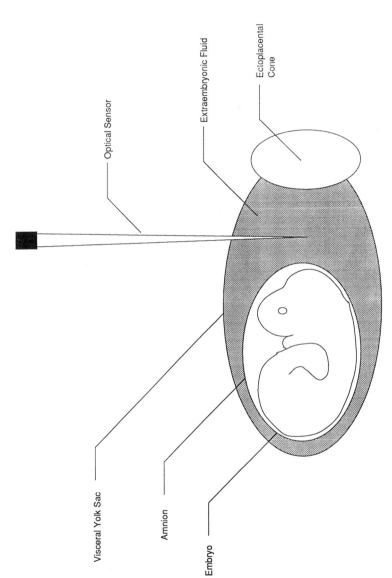

Figure 2. Schematic representation of the rat conceptus with its associated tissues, showing placement of the ultramicrofiberoptic sensor inside the extraembryonic fluid compartment.

Optical Sensor

Extraembryonic Fluid

Ectoplacental Cone

Visceral Yolk Sac

Amnion

Embryo

The miniaturization of this fiber optical sensor has not occurred at the expense of response times, with current sensors able to respond to events of shorter than 100ms (*12*). We made use of this feature in determining the dynamic response of a rat conceptus to alterations in its *in vitro* environment. Experiments were conducted to ascertain the ability of the GD 12 rat conceptus to respond to changes in its environment and to selected chemical agents. Three different experimental protocols were used. First, the perifusate bathing the embryo was saturated with 100% nitrogen, to create a condition of hypoxia. The measured pH over time did not vary significantly from the initial control readings. Even though hypoxia does not immediately alter hydrogen ion concentrations in the conceptus as determined by measuring EEF pH, it does not preclude the possibility that intracellular changes are taking place within cells of the VYS or embryo proper, but not affecting the equilibrium between EEF and intracellular fluid compartments. Second, conceptal response was determined by monitoring EEF for pH, when the perifusate, containing HBSS, was altered over a pH range of 6.6 to 8.6. Again, the measured pH in EEF did not result in any significant variations from control levels. Finally, the dynamic effects of a thiol oxidant, diamide (used to oxidize intracellular glutathione (GSH) to glutathione disulfide (GSSG) and induce oxidative stress in the embryo) was evaluated. As shown in Figure 3, a 500 μM solution of diamide in the perifusate (HBSS, pH 7.4) results in an initial rapid decrease in pH over the first 30 seconds, followed by a slower downward trend thereafter. An absolute decrease of about 0.3 pH units occurs within about 3 minutes.

In conclusion, the ultramicrofiberoptic pH sensor is shown to be able to discriminate pH changes of less than a tenth of a pH unit over the range of one pH unit above and below the physiologic pH of 7.4. The pH measurements were obtained in real time, on a single, intact, viable rat conceptus under conditions of environmental change and direct chemical exposure. The insertion of the ultrasmall sensor through the visceral yolk sac into the EEF appeared to cause no damage to or leakage from the involved tissues. This demonstrates the advantage of an essentially non-invasive approach, as compared to conventional means, necessitating the disruption of large numbers of conceptuses to obtain less sensitive and indirect measurements. Conceptuses of ascending developmental age undergo an acidification of the EEF. This finding is consistent with reports using accumulation of ^{14}C-dimethadione (5,5'-dimethyloxazolidine-2,4-dione) or DMO, to determine tissue pH (*17,18*). Alterations in environmental conditions showed the ability of the rat conceptus to respond to certain external stimuli by maintaining normal intraconceptal pH. On the other hand, diamide, at a concentration of 500 μM, produced a very drastic decrease in pH over the 30 seconds of chemical exposure. This thiol oxidant may compromise the normal buffering capacity of rat conceptuses through disruption of energy dependent ion compartmentalization or inhibition of H^+ pumps. The observed pH changes (~0.3 pH units) are of a magnitude capable of dramatically altering normal cell function, as has been shown in several other cell systems (*19*). This observation is important, considering the evidence that changes in intracellular pH may be instrumental in the molecular regulation of differentiation and cell proliferation during development. Intracellular thiol status has also been implicated in the regulation of the same developmental processes. Biological hydrogen ion pumps are, in fact, known to be particularly sensitive to disturbances in intracellular thiol status. The lysosomal ATP-dependent H^+ pump is selectively inhibited by the sulfhydryl reagent N-ethylmaleimide (NEM), at doses that do not affect other ion pumps (*20*). The ability of ultramicrofiberoptic sensors to measure pH changes, in real time, in the intact rat conceptus, demonstrates their potential applications for dynamic analysis in small multicelluar organisms. Working sensor probe dimensions and response characteristics also make this approach feasible for use in single cells. The application of this novel technology to studies of developmental regulation, pharmacokinetics,

TABLE 1

Measurements of pH in the extraembryonic fluid (EEF) of whole rat conceptuses maintained in vitro as determined using the micro-fiberoptic pH sensor.

Day of Gestation	Somite Number	pH
10	10-16	7.53 ± 0.07
11	22-26	7.37 ± 0.07
12	32-34	7.26 ± 0.06

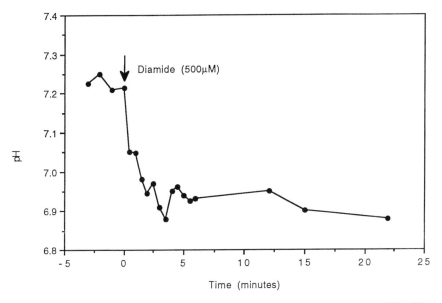

Figure 3. Representative dynamic pH response during exposure to 500 μM diamide, determined in the extraembryonic fluid of single, viable, GD 12 rat conceptus, using the ultramicrofiberoptic pH sensor. Diamide is added at t=0 and the decrease in pH during the first 30 seconds was determined to be 0.18±0.01 pH units (n=3 conceptuses).

272

toxicology and physiology will provide valuable spatial and temporal information not heretofore available using conventional techniques.

Acknowledgments

Supported by NIH grants 1RO1GM50300 and ES05235and by a grant from the Physical Optics Corporation, Torrance, CA.

Literature Cited

1. Seitz, W. R. *Anal. Chem.* **1984**, *56*, 16A; CRC Crit. Rev. *Anal. Chem.*, **1988**, *19*, 135.
2. Barnard, S. M.; Walt, D. R. *Nature* **1991**, *353*, 338.
3. Meyerhoff, M. E.; Opdycke, W. N. In *Advances in Clinical Chemistry*; Spiegel, H. E., Ed.; Academic Press, Inc.: 1986, Vol. 25, pp 1-47.
4. Wightman, R. M.; Wipf, D. O. In *Electroanalytical Chemistry*; Bard, A. J., Ed.; Dekker: New York, NY, 1989, Vol.15; p 267.
5. Wightman, R. M. *Science* **1988**, *240*, 415.
6. Strein, T. G.; Ewing, A. G., *Anal. Chem.* **1992**, *64*, 1368.
7. Ji, S.; Lemasters, J. J.; Thurman, R. G. *FEBS Lett.* **1980**, *113*, 37.
8. Thurman, R. G.; Lemasters, J. J. *Drug Metab. Rev.* **1988**, *19 (3,4)*, 263.
9. Herman, B.; Jacobson, K. *Optical Microscopy for Biology*; Wiley Liss: New York, NY, 1989.
10. Nau, H.; Scott, W. J. *Arch. Toxicol.* **1987**, *Suppl. 11*, 128 -139.
11. Tan, W.; Shi, Z -Y; Kopelman, R. *Anal. Chem.* **1992**, *64*, 2985-2990.
12. Tan, W.; Shi, Z -Y; Smith, S.; Birnbaum, D; Kopelman, R. *Science* **1992**, *258*, 778-781.
13. Tan, W.; Kopelman, R. In *Fluorescence Imaging Spectroscopy and Microscopy*; Wang, X. F., Herman, B., Eds; John Wiley & Sons, Inc.: New York, NY, 1996, pp 407-475.
14. Betzig, E.; Troutman, J. K.; Harris,T. D.; Weiner, J. S.; Kostelak, R. L.; *Science* **1991**, *251*, 1468; Betzig, E; Troutman, J. K. *Science* **1992**, *257*, 189.
15. Harris, C.; Juchau, M. R.; Mirkes, P. E. *Teratology* **1991**, *43*, 229.
16. Freeman, S. J.; Coakley, M.; Brown, N. A. In *Biochemical Toxicology: A Practical Approach*; Snell, K., Mullock, B., Eds.; IRL Press, 1987, pp 83-107.
17. Nau, H.; Scott, W. J. *Nature* **1986**, *323*, 276.
18. Scott, W. J.; Duggan, C. A.; Schreiner, C. M.; Collins, M. D. *Toxicol. Appl. Pharmacol.* **1990**, *103*, 238.
19. Busa, W. B. *Ann. Rev. Physiol.* **1986**, *48*, 389-402.
20. Dean, G. E.; Fishkes, H.; Nelson, P. J.; Rudnick, G. *J. Biol. Chem.* **1984**, *259*, 9569.

Chapter 23

The Use of Optical-Imaging Fibers for the Fabrication of Array Sensors

Karri L. Michael, Jane A. Ferguson, Brian G. Healey, Anna A. Panova, Paul Pantano, and David R. Walt

Max Tishler Laboratory for Organic Chemistry, Department of Chemistry, Tufts University, Medford, MA 02155

We have used coherent imaging fibers to make fiber-optic chemical sensors. Two main approaches have been taken. In the first approach, sensors were made with spatially-discrete sensing sites for multianalyte determinations. In the second approach, imaging fibers were used to fabricate array sensors that can concurrently view a sample and detect a single analyte. Such sensors have near micron resolution and can monitor chemical changes occurring in tens of milliseconds.

An optical fiber is a filamentous, dielectric material that conducts light from one end to the other. The fiber is composed of an inner core material surrounded by an outer cladding material with a lower refractive index. Light entering within a certain critical angle is totally internally reflected at the interface between these two materials, thereby guiding light through the optical fiber. Sensors are fabricated by placing a polymer layer containing a covalently immobilized fluorescent indicator on the fiber's distal end. The indicator's optical properties are monitored as they are modulated by the presence of an analyte.

In most situations where measurements are required, it is important to monitor several analytes simultaneously. Traditionally, multianalyte optical sensing has been accomplished by using several different indicators, each on its own optical fiber, that exhibit some preferential response to a particular analyte. This approach avoids the problem of signal mixing due to the broad spectral signature of most indicators. Recently, a single sol-gel clad fiber-optic waveguide was used for multianalyte sensing (1). In this approach, the fiber's cladding was removed in 4-cm sections at several locations and the fiber was dip-coated at each location with a different indicator-doped sol-gel solution. Each indicator was spatially resolved by pulsed evanescent excitation and time-resolved detection. Spatial resolution of the indicators was determined by using the fluorescent lifetime.

In our laboratory, multianalyte sensors are fabricated by spatially resolving the indicators using a coherent imaging fiber. An imaging fiber, comprising thousands (*ca.*~6000) of individual 2-4-μm diameter optical fibers, each with its own core and cladding, is shown schematically in Figure 1. Individual fibers are melted and drawn together in a coherent manner such that each fiber carries an

[1]Corresponding author.

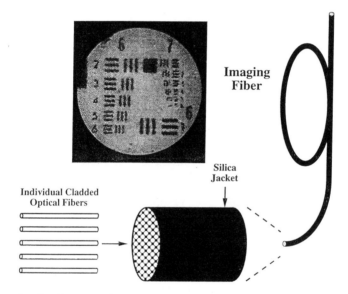

Figure 1. A coherent optical imaging fiber. (Bottom) An imaging fiber is a flexible fiber array comprised of thousands of individually-cladded optical fibers that are melted and drawn together such that an image can be carried from one face to the other. (Top) Image of an Air Force resolution target acquired with a CCD camera, as viewed through a 350-μm-diameter AChE/PAN/FITC-modified imaging fiber.

isolated optical signal allowing images to be carried and maintained from one end to the other (Figure 1). Discrete sensing regions are immobilized in precise locations on the fiber's distal face to create a multianalyte sensor. Combining the imaging fiber with a charge-coupled device (CCD) camera allows simultaneous, independent measurements of numerous analytes, with 2-4-μm spatial resolution (limited essentially by the size of the individual fiber elements in the array). As examples of this approach, we present results from a pH, CO_2 and O_2 multianalyte sensor; a penicillin/pH biosensor; and DNA sensor arrays capable of detecting single-base mutations.

In a second approach, the imaging fiber's individual optical pathways are used to concurrently view a sample and detect a single analyte by placing a uniform, planar sensing layer on the fiber's distal face. Preliminary *in situ* results are presented whereby an acetylcholine-sensitive imaging fiber is used to monitor neurochemical dynamics from cells and tissue. In a very different application, results are also presented from a pH-sensitive imaging fiber used to measure corrosion processes occurring on metal surfaces.

Multianalyte Sensors

Optical imaging fibers are excellent substrates for multianalyte sensing. Multianalyte sensors are fabricated by site-selective photodeposition of analyte-sensitive polymer matrices on an optical imaging fiber's distal face (2-6). The analyte-sensitive polymer matrices are photodeposited using collimated light from a mercury-xenon arc lamp. The collimated light is passed through neutral density and excitation filters, focused onto a pinhole and imaged onto the proximal end of an imaging fiber using a 15x reflective microscope objective. The light travels through the discretely-illuminated portion of the imaging fiber (20-80 μm) to the distal face, which is placed in a photopolymerization solution (consisting of monomer, initiator and indicator). The thickness of each analyte-sensitive polymer matrix is determined by controlling the illumination time using an electronic shutter. Using this photodeposition method, multiple analyte-sensitive polymer matrices are deposited spatially by repositioning the imaging fiber and repeating the process using either the same or different photopolymerization solutions.

The instrument used for fluorescence measurements is a modified epifluorescence microscope (7). White light from a xenon arc lamp is collimated, passed through the appropriate excitation filter and focused onto the imaging fiber's proximal face using a 10x or 20x microscope objective. The excitation light travels to the distal end of the fiber where the analyte-sensitive polymer matrices fluoresce in proportion to the analyte concentrations. The fluorescence returning to the proximal end of the imaging fiber is collected by the microscope objective, transmitted through a dichroic mirror, filtered at the appropriate emission wavelength, and detected by a CCD camera. In operation, this instrument obtains continuous ratiometric measurements of various analytes by interchanging excitation and emission filters (at a rate of 20 Hz) using computer-controlled filter wheels and shutters.

Simultaneous Monitoring of pH, CO_2 and O_2. Multianalyte sensing using imaging fibers has been demonstrated for simultaneous detection of pH, CO_2 and O_2. The distal face of a 250-μm diameter optical imaging fiber was polished, cleaned, and silanized with 3-(trimethoxysilyl)propyl methacrylate. Silanization provides the fiber surface with polymerizable groups that facilitate covalent attachment of acrylic polymers. First, two 40-μm diameter oxygen-sensitive spots were deposited by entrapping a ruthenium derivative ($Ru(Ph_2phen)_3^{2+}$) in a siloxane polymer containing pendant acrylate groups. Second, four 40-μm diameter pH-sensitive regions were deposited on the fiber by covalently immobilizing acryloylfluorescein in a poly(hydroxyethylmethacrylate) (polyHEMA) hydrogel. Finally, two of the pH spots were converted into CO_2-sensitive regions by saturating the polyHEMA hydrogel with a CO_2-sensitive bicarbonate buffer and subsequently overcoating with a gas-permeable hydrophobic siloxane polymer membrane.

The sensor was calibrated by simultaneously measuring the change in fluorescence intensity of each analyte-sensitive matrix when immersed in solutions of known pH, or in solutions equilibrated with calibrated gases (3). The fluorescence of the O_2-sensitive matrix decreases in the presence of oxygen due to quenching of the ruthenium dye. The pH-sensitive matrix exhibits high fluorescence under basic conditions and low fluorescence under acidic conditions. When CO_2 crosses the gas-permeable siloxane membrane, carbonic acid is generated decreasing the pH of the buffer behind the membrane. Therefore, the CO_2-sensitive matrix decreases in fluorescence intensity in the presence of CO_2; a pH change in the bulk solution does not affect the CO_2-sensitive matrix. The multianalyte sensor has dynamic ranges of 0-100% for oxygen, 0-10% for carbon dioxide, and 5.5-7.5 for pH (6).

The multianalyte *in situ* measurement capability was demonstrated in a beer fermentation application. The sensor was stabilized initially in unfermented wort saturated with various levels of CO_2 and O_2 (Figure 2). Yeast was then added to the wort and the sensor simultaneously measured the pH, CO_2 and O_2 changes during the course of fermentation. When yeast is added to unfermented wort, CO_2 is produced by the partial oxidation of glucose (8). This production of CO_2 causes the pH of the wort to decrease. At the same time, oxygen in the medium is consumed initially by the yeast for the synthesis of lipids in the yeast cell membrane (9). The calibration data were converted to concentration versus intensity plots, which enabled the fermentation data to be graphed as time versus concentration for each analyte (data not shown). Although commercially available oxygen analyzers were not used for comparison during the fermentation reactions, several have been used in experiments (e.g., blood gas analysis and bioremediation monitoring) yielding comparable results.

Simultaneous, Independent Measurement of pH and Penicillin. Biosensors based on the multianalyte array sensor concept have also been demonstrated. Most enzyme-based biosensors operate by coupling the enzymatic activity to a measurable parameter such as pH, O_2 or H_2O_2. Such biosensors show excellent selectivity for the analyte of interest and are highly sensitive however, they suffer from errors due to changes in the concentration of the chemical species involved in the transduction mechanism. We have developed a multianalyte biosensor design for the simultaneous and independent measurement of both the analyte of interest (penicillin) and the transducing chemical species (hydrogen ions) (5). A dual-analyte penicillin biosensor was fabricated by photodepositing an array of pH-sensitive polymer matrices and penicillin-sensitive polymer matrices on a single imaging fiber. The response of the penicillin-sensitive polymer matrices is based on a microenvironmental pH change in the polymer as penicillin is converted to penicilloic acid by penicillinase. A pH-sensitive indicator is coimmobilized within the polymer matrix to measure the polymer's microenvironmental pH. A separate pH sensor on the imaging fiber measures the sample pH. The differential response of the dual-analyte biosensor allows for the pH-independent measurement of penicillin.

The array was fabricated using the photodeposition technique described above. Four 27-μm diameter pH-sensitive polymer matrices were deposited by immobilizing acryloylfluorescein in a cross-linked polyHEMA hydrogel. Second,

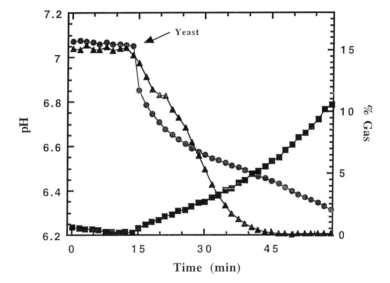

Figure 2. Simultaneous, *in situ* measurements of pH (circles), CO_2 (squares) and O_2 (triangles) during a beer fermentation using a single 250-μm-diameter optical imaging fiber.

four penicillin-sensitive polymer matrices were deposited by photopolymerizing a HEMA monomer solution containing both acryloylfluorescein and a lyophilized penicillinase suspension. After fabrication, the immobilized penicillinase was activated by hydrating the biosensor in pH 7 phosphate buffer. Fluorescence images of the biosensor's response to 0 mM and 5 mM penicillin in pH 7 phosphate buffer are shown in Figure 3 (top and bottom, respectively). In the presence of penicillin, the penicillin-sensitive matrices decrease in fluorescence intensity due to the increased microenvironmental acidity from the enzymatic hydrolysis of penicillin; the pH-sensitive matrices show no change in fluorescence intensity in the presence of penicillin.

In order to test the biosensor's ability to measure penicillin, a penicillin batch fermentation using *Penicillium chrysogenum* was performed. This fermentation was a challenging test for the biosensor because the medium pH decreases as penicillin is produced. During the fermentation, samples were periodically withdrawn and stored at -20°C until the fermentation was complete. When the fermentation was complete, the samples were brought to room temperature and tested using a calibrated penicillin biosensor and a colorimetric reference method (10-11). Figure 3 shows how the sensor response compares with the colorimetric method versus time. The figure also shows the fermentation pH as measured by the pH matrices. These results demonstrate that the penicillin biosensor and the reference method are in excellent agreement and that the penicillin biosensor can accurately measure the penicillin concentration even with concomitant pH changes.

DNA Microarrays. The multianalyte sensor concept has also been used to fabricate optical DNA biosensors (12-13). These sensors are used to monitor the hybridization of fluorescently-labelled target oligonucleotides to complementary probes immobilized on the optical fiber. Two sensor array designs were utilized. The first array design, shown in Figure 4, comprised seven single core fibers bundled together at their proximal ends. The fibers' distal ends were silanized and functionalized independently with different cyanuric chloride-activated oligonucleotide probes (14). Probes specific for human cytokine mRNA sequences (e.g., interleukin 6 [IL6], IL4, human β-globin, interferon γ-1 and IL2) were chosen due to the widespread interest in cytokine gene expression in response to inflammatory stimuli or infection (12).

Sensor specificity was determined by assaying solutions containing fluorescein-labelled targets complementary to one or more of the probes. Specific hybridization was observed as a fluorescent signal developing only on fibers containing probes complementary to the solution targets (Figure 4). Non-complementary labelled targets resulted in a complete absence of signal, confirming hybridization specificity. Unlabelled samples can also be assayed in competition with fixed amounts of labelled target. The unlabelled target hybridizes in proportion to its concentration enabling quantitative, as well as qualitative, measurements. Using either method, sensor regeneration can be accomplished by dipping the array in pH 8.3 TE buffer (10 mM Tris-HCl, 1 mM EDTA) containing 90% formamide. Subsequent analyses using the same sensor showed comparable signals with complementary probe/target pairs. This approach enables fast (<10 min) and sensitive (10 nM) detection of multiple DNA sequences simultaneously.

The second DNA biosensor design involved covalently attaching 5'-amino terminal DNA to acrylamide-based polymer matrices photodeposited directly on an imaging fiber (13). An array was prepared containing two oligonucleotide probes (A and A*), differing by one base pair only (Figure 5). In brief, two acrylamide and N-acryloxysuccinimide copolymer matrices were photodeposited on a silanized imaging fiber. The fiber was then placed in a solution of 5'-amino terminated oligonucleotide A*, covalently immobilizing the oligonucleotide via the polymer's

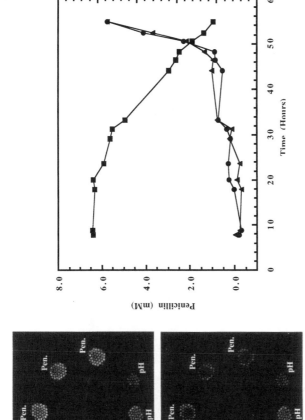

Figure 3. (Left) Fluorescence images (490 nm excitation, 530 nm emission) of a penicillin/pH array in pH 7 phosphate buffer containing 0 mM (top) and 5 mM (bottom) penicillin, respectively. (Right) Graph of the pH change (squares) and the penicillin concentration during the fermentation as determined by both the penicillin/pH array (triangles) and a colorimetric reference method (circles).

280

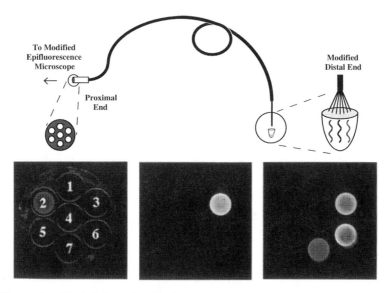

Figure 4. (Top) Schematic of a DNA biosensor array comprised of seven single core fibers bundled together at their proximal ends. Each single core fiber was modified with a different oligonucleotide probe. (Bottom) Fluorescence images of the fiber bundle when exposed to fluorescein-labelled, complementary targets. The images show specific hybridization of the complementary targets (white) in samples containing single targets (left, middle) and multiple targets (right).

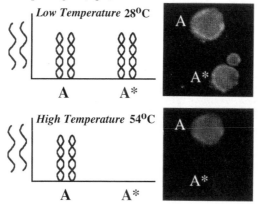

Figure 5. Fluorescence images of the DNA biosensor hybridization to a FITC-labelled oligonucleotide A target at 28 °C (Top) and 54 °C (Bottom). The images show the sensor's ability to distinguish single base-pair mutations (A versus A*) at higher temperatures.

reactive succinimidyl ester residues. Any unreacted ester residues were treated with ethanolamine to ensure the immobilization of only one oligonucleotide sequence per polymer matrix. A second copolymer matrix was deposited and allowed to react with oligonucleotide A.

The biosensor was used to discriminate single base pair mismatches (A* and A) by monitoring the real-time hybridization of an FITC-labelled target oligonucleotide complementary to A. As shown in Figure 5, the target hybridizes to both sequences at 28°C. At higher temperatures (e.g., 54°C) hybridization occurs only to the perfect complement, oligonucleotide A, and not to the single base pair mismatch, oligonucleotide A*. Using this method, the DNA biosensor array could detect FITC-labelled target oligonucleotide in the range 2-200 nM. The biosensor could be regenerated in < 10 s by placing the hybridized biosensor in a 2X SSPE (20X SSPE contains 3 M NaCl, 0.2 M NaH_2PO_4 and 0.02 M EDTA pH 7.4), 0.01% SDS buffer at 65°C. Signal levels remained unchanged after three sensor regenerations.

Combined Imaging and Chemical Sensing

Another use of these imaging fibers and chemically-sensitive polymer matrices combines imaging with chemical sensing. This technique provides the ability to both view a sample and measure chemical dynamics occurring at the sample surface (e.g., neurotransmitter release or metal dissolution). Such sensors are prepared by covalently immobilizing a uniform, planar sensing layer on the entire distal face of the imaging fiber using spin-coating techniques. Thin polymer layers (≤ 5 μm), do not hinder the fiber's imaging capabilities (Figure 1). Using a modified epifluorescence microscope, a CCD camera and a single imaging fiber, transmitted images can be acquired for visual and morphological information which can be coupled with fluorescence images of local chemical concentrations (Figure 6). Thus, in a single fabrication step, a coherent sensing array of thousands of microsensors can be integrated and read simultaneously, with 2-4 μm spatial accuracy. The fluorescence images are acquired as described previously for multianalyte sensors; transmitted images are obtained by externally illuminating the sample, and viewed through a neutral density filter.

These modified imaging fibers were developed as remote sensing tools to eliminate the need to bring the sample to the microscope stage. Such arrays also eliminate the need to load the sample with indicators, do not invade or penetrate the sample, do not require a high degree of precision to position (as compared to extremely small probes) and offer major advantages over microelectrode arrays in both ease of fabrication and manipulation at the cellular level. Biological samples imaged with these planar, polymer-modified imaging fibers have included mouse fibroblast cells (10-20-μm diameter), sea urchin eggs (50-100-μm diameter) and cancerous human lymph node tissue slices. The versatility of the combined imaging and chemical sensing technique has been further demonstrated in systems as diverse as neuronal preparations and metal surfaces.

Acetylcholine (ACh) Biosensor. We have reported previously an acetylcholine-sensitive biosensor with a detection limit of 35 μM and a subsecond response time (7). Briefly, a 350-μm diameter imaging fiber containing ~ 6000 optical fibers was polished and the distal face was amino-silanized to activate the fiber's surface. An ACh-sensitive polymer layer was created by co-immobilizing acetylcholinesterase (AChE) in a water-soluble, functionalized pre-polymer known as poly(acrylamide-co-N-acryloxysuccinimide) (PAN) (15). The AChE-derivatized PAN was spin-coated onto the amino-silanized fiber surface and allowed to cure. The AChE/PAN-modified imaging fiber was then allowed to react with the pH-sensitive indicator fluorescein isothiocyanate (FITC). When ACh is introduced to the AChE/FITC-modified polymer layer, it is hydrolyzed to choline and acetic acid by the immobilized enzyme. The dissociated protons from the enzyme-generated acetic acid then quench the immobilized FITC to produce a decrease in fluorescence intensity proportional to the ACh concentration (Figure 7).

While AChE hydrolyzes various esters of acetic acid (16), the selectivity of an AChE/FITC/PAN-modified imaging fiber to ACh was shown by its lack of response to butyrylcholine. This lack of response is due to the difference in AChE activity towards these two substrates, and not a lack of sensitivity of the modified imaging fibers to enzyme-generated butyric acid, since the sensors were shown to be equally responsive to both acetic acid and butyric acid. Finally, a FITC/PAN-modified imaging fiber containing no AChE gave no response to ACh, demonstrating that the ACh response is enzyme-generated (7).

In association with Professor Guillermo R. Pilar and Carlos Pena (School of Medicine, Case Western Reserve University), we are currently employing the acetylcholine-sensitive biosensor to measure the stimulated release of acetylcholine from cell surfaces. In this application, the biosensor is brought into contact simultaneously with the surface of hundreds of single dissociated ciliary ganglion neurons (10-20-μm diameter) from Stage 34-35 white Leghorn chick embryos. In these experiments, fluorescence images are acquired continuously before, during and after the delivery of the calcium ionophore A23187, a neurochemical stimulant, using a pressure-ejection system. We observe a fluorescence response only in the areas where ciliary ganglion neurons contact the ACh-sensitive fiber; areas not in contact with a ganglion surface show no response (Figure 8). In addition, the fluorescence response is dependent upon the addition of the neurochemical stimulant; delivery of buffer shows no response. With this technique, white light images of the cells can be used in conjunction with the fluorescent images to correlate morphological information with chemical information. Both types of information are valuable, since a neuron's shape, and the number and function of its processes, can distinguish one neuron type from another. For this application, combining imaging and sensing has the advantage of measuring neurotransmitter release from hundreds of cells simultaneously, not only providing information on a cell-by-cell basis, but also providing a way to map cellular communication and connectivity.

In association with Professor Barry A. Trimmer (Department of Biology, Tufts University), the ACh biosensor was also demonstrated in a sensory ganglion preparation from the larval tobacco hornworm, *Manduca sexta*. In these experiments, fluorescence images were acquired prior to, during and after the sensory nerve of the ganglion was stimulated electrically via a suction electrode. We have observed a fluorescence response that mirrors the electrical stimulus and is regionally differentiated within the ganglion. In this application, white light images provide information about spatial organization as well as function localization. We are currently performing biological controls to unequivocally demonstrate that the responses observed in both the chick ciliary ganglia neurons and the hornworm ganglion are cholinergic in nature.

Imaging and Sensing of Corrosion Processes. In addition to biological samples, the combined imaging and chemical sensing technique has shown its usefulness in studying corrosion processes at the microscopic level. Using the same spin-coating technique described above, a pH-sensitive, PAN/FITC-modified imaging fiber was fabricated to study the onset of corrosion. The pH-sensitive imaging fiber's response was calibrated in 5 mM phosphate buffers of varying pH containing 0.1 M KCl. The calibration curve shows the typical sigmoidal shape of a pH titration curve (Figure 9A).

The ability to measure pH changes from metal surfaces was first evaluated under electrochemically-generated corrosion conditions employing a

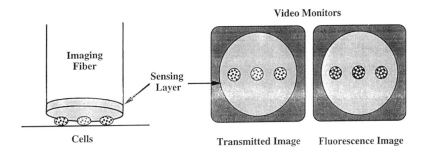

Figure 6. Principle of the combined imaging and chemical sensing technique. The technique provides the ability to both view a sample (transmitted image) and measure surface chemical dynamics (fluorescence image) using a single optical imaging fiber.

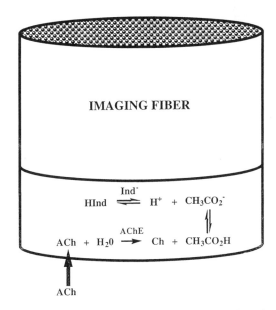

Figure 7. Principle of an AChE/PAN/FITC-modified imaging fiber. In operation, AChE catalyzes the hydrolysis of ACh to choline and acetic acid; the dissociated protons from the enzyme-generated acetic acid react with the immobilized FITC to provide a fluorescence signal proportional to the ACh concentration.

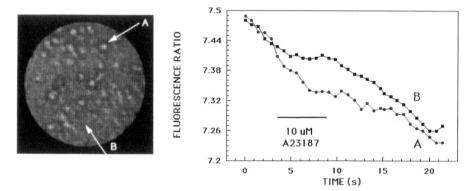

Figure 8. Preliminary biological measurements of an AChE-modified imaging fiber held in contact with dissociated ciliary ganglion neurons from white Leghorn chick embryos. (Left) White light image of dissociated ciliary ganglion neurons as viewed through the biosensor. In both the image and the graph, A) is a region of the biosensor in contact with a ciliary ganglion neuron; B) is a region of the biosensor not in contact with a ciliary ganglion neuron. The graph shows the biosensor's response upon stimulation with the calcium ionophore A23178. Measurements were acquired with a 200-ms CCD acquisition time.

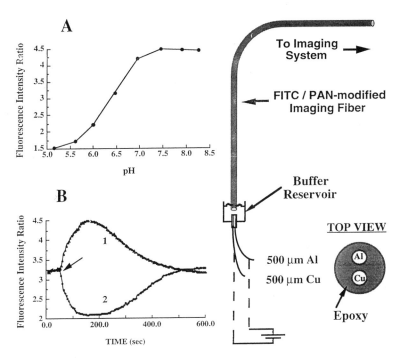

Figure 9. Experimental setup used for monitoring pH changes on metal surfaces with a pH-sensitive optical imaging fiber. A) Calibration curve of a FITC/PAN-modified imaging fiber. B). The ratio of mean fluorescence (490/440 nm) at 530 nm acquired with a FITC/PAN-modified imaging fiber held in contact with (1) a copper electrode and (2) and aluminum electrode during the application (shown by arrow) of (1) - 0.2 V, relative to the aluminum electrode and (2) + 0.2 V, relative to the copper electrode.

copper/aluminum electrochemical cell. The cell was constructed by embedding both an aluminum and a copper wire (*ea.* 500-μm diameter) in an epoxy resin, polishing one end, and placing the cell facing upward in a vial containing 5 mM phosphate buffer (pH 6.5, 0.1 M KCl) (Figure 9). The pH-sensitive imaging fiber was brought into contact with the copper wire and a constant cathodic potential (-0.2 V relative to the aluminum electrode) was applied for 75 s while acquiring a sequence of fluorescence ratio pairs (ex. 490/440 nm, em. 530 nm). The generation of hydroxide ions, due to the reduction of water (Equation 1) and/or the reduction of oxygen (Equation 2) (*17*), resulted in a localized increase in fluorescence

$$2\,H_2O \ + \ 2\,e^- \ = \ 2\,OH^- \ + \ H_2$$

(1)

$$O_2 \ + \ 2\,H_2O \ + \ 4\,e^- \ = \ 4\,OH^-$$

(2)

intensity (Figure 9B). After the potential was removed, the sensor gradually returned to its initial intensity as equilibrium was reestablished in the polymeric sensing layer. Conversely, the pH-sensitive imaging fiber was placed over the aluminum surface and an anodic potential (+0.2 V relative to the copper electrode) was applied for 75 s during the acquisition of fluorescence ratio pairs. The dissolution and solvolysis of aluminum (Equations 3 and 4) (*17*)

$$Al \ = \ Al^{3+} \ + \ 3\,e^-$$

(3)

$$Al^{3+} \ + \ 3\,H_2O \ = \ Al(OH)_3 \ + \ 3\,H^+$$

(4)

resulted in proton evolution leading to a localized decrease in fluorescence intensity (Figure 9B). After the potential was removed, the sensor gradually returned to its initial intensity as equilibrium was reestablished. These results demonstrate the pH-sensitive imaging fiber's ability to distinguish between base-generating cathodic and acid-generating anodic sites.

Two dissimilar metals will develop a potential difference when immersed in a conductive buffer solution. This two-metal, or galvanic, corrosion could be investigated using a copper-plated aluminum wire. The copper-plated wire was fabricated by depositing a 10 mM $CuSO_4$ solution containing 5mM KCl on the surface of a polished aluminum wire. An aluminum-copper corrosion cell was then made by scratching a region of the copper layer several micrometers wide to expose the aluminum. A PAN/FITC-modified imaging fiber was placed over the scratched-wire surface to measure the pH changes (Equations 1, 2, 4) occurring as a result of electron flow between the two metals upon immersion into a corrosive solution (5 mM phosphate buffer, 0.1 M KCl, pH 6.5). Localized regions of both high and low fluorescence intensities developed on the wire surface near the scratch (Figure 10). In this region, the aluminum was exposed to the corrosive medium, resulting in aluminum dissolution (Equations 3 and 4). Aluminum dissolution was facilitated by the presence of the adjacent copper layer where the corresponding reduction reactions (Equations 1 and 2) occur more readily due to the intrinsic potential difference between the two metals. Intensive aluminum dissolution is shown by the fluorescence images in Figure 10 A-F (low fluorescence regions, denoted by black, are low pH regions). After 30 s of exposure to the corrosive solution (Figure 10 A), the reactive scratch was ~ 6-μm wide and expanded to ~ 15-μm wide as passive aluminum regions became active corrosion sites. After 5 min of exposure, the active

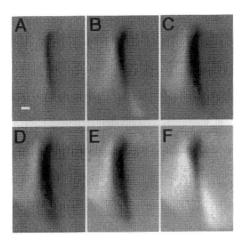

Figure 10. Background-subtracted fluorescence images (relative to fluorescence at 2 sec) acquired with an FITC/PAN-modified imaging fiber. The fiber was placed over a scratch made in a copper-plated aluminum wire and images were acquired after (A) 30 s, (B) 2 min, (C) 4 min, (D) 5 min, (E) 6 min and (F) 7 min of exposure to pH 6.5 phosphate buffer containing 0.1 M KCl. High fluorescence intensities are denoted by white. The scale bar in (A) is 10 μm and is the same for all images.

aluminum sites began to passivate as a salt layer formed. Regions of high fluorescence intensity (denoted by white) may be due to an increase in hydroxide evolution as a result of cathodic reactions on the copper surface and/or a decrease in proton evolution as the aluminum surface was passivated.

Determining the absolute pH values during the corrosion processes, together with monitoring initial stages of localized forms of corrosion, such as pitting and crevice corrosion, are the subjects of future research. At the present stage it can be concluded that the technique is capable of real-time chemical sensing of corrosion sites as small as several microns with spatial resolution limited by both the diameter of the individual optical fibers comprising the imaging fiber and by diffusion. The technique can be applied to other surface phenomena as well as sensing chemical species other than hydrogen and hydroxide ions.

Conclusions

We have demonstrated the widespread utility of coherent imaging fibers for fabricating sensor arrays. Sensors were made with spatially-discrete sensing sites for multianalyte determinations. These sensors solve many of the problems associated with designing optical multianalyte sensors, such as spectral overlap of multiple indicators and the need to use individual optical fibers for each analyte. Imaging fibers were also used to fabricate array sensors that can concurrently view a sample and detect a single analyte. These sensors can display visual information of a remote sample with 2-4-μm spatial resolution, allowing for alternate acquisition of both chemical and visual characteristics.

Acknowledgments

We gratefully acknowledge the National Institutes of Health (GM-48142) for financial support. We also thank Dr. MaryBeth Tabacco for editing assistance, Dr. K. Sue Cook (Tufts University, Department of Biology) for providing mouse fibroblast cells and Dr. Susan Ernst (Tufts University, Department of Biology) for providing sea urchin eggs.

Literature Cited

1. Browne, C. A.; Tarrant, D. H.; Olteanu, M. S.; Mullens, J. W.; Chronister, E. L. *Anal. Chem.* **1996**, *68*, 2289.
2. Walt, D. R.; Agayn, V.; Healey, B. G. In *Immunoanalysis of Agrochemicals: Emerging Technologies*. Nelson, J. O.; Karu, A. E.;Wong, R. B., Eds.; ACS Symposium Series 586; American Chemical Society: Washington, DC, 1995; pp 186-196.
3. Pantano, P.; Walt, D. R. *Anal. Chem.* **1995**, *67*, 481A.
4. Li, L.; Walt, D. R. *Anal. Chem.* **1995**, *67*, 3746.
5. Healey, B. G.; Walt, D. R. *Anal. Chem.* **1995**, *67*, 4471.
6. Ferguson, J. A.; Healey, B. G.; Bronk, K. S.; Barnard, S. M.; Walt, D. R. *Analytica Chimica Acta.* **1997**, *340*, 123.
7. Bronk, K. S.; Michael, K. L.; Pantano, P.; Walt, D. R. *Anal. Chem.* **1995**, *67*, 2750.
8. Uttamlal, M.; Walt, D. R. *Bio/Technology.* **1995**, *13*, 597.
9. Kirshop, B. H.; *J. Inst. Brew.* **1974**, *80*, 252.
10. Nilsson, H.; Mosbach, K.; Enfors, S.; Molin, N. *Biotechnol. Bioeng.* **1987**, *20*, 527.
11. Boxer, G. E.; Everett, P. M. *Anal. Chem.* **1949**, *21*, 670.
12. Ferguson, J. A.; Boles, T. C.; Adams, C. P.;Walt, D. R. *Nature Biotech.* **1996**, *14 (13)*, 1681.

13. Healey, B. G.; Matson, R. S.; Walt, D. R. *Analytical Biochem.* **1997**, in press.
14. Van Ness, J.; Kalbfleisch, S.; Petrie, C. R.; Reed, M. W.; Tabone, J. C.; Vermeulen, N.M.J. *Nucleic Acids Res.* **1991**, *19*, 3345-3350.
15. Pollak, A.; Blumenfeld, H.; Wax, M.; Baughn, R. L.; Whitesides, G. M. *J. Am. Chem. Soc.* **1980**, *102*, 6324.
16. Froede, H. C.; Wilson, I. B., In *The Enzymes*; Boyer, P. D., Ed.; Academic Press: New York, NY, 1971; Vol. 5, pp 87-114.
17. Engstrom, R. C.; Ghaffari, S.; Qu, H. *Anal. Chem.* **1992**, *64*, 2525.

INDEXES

Author Index

Subject Index

Bestsellers from ACS Books

The ACS Style Guide: A Manual for Authors and Editors (2nd Edition)
Edited by Janet S. Dodd
470 pp; clothbound ISBN 0–8412–3461–2; paperback ISBN 0–8412–3462–0

Writing the Laboratory Notebook
By Howard M. Kanare
145 pp; clothbound ISBN 0–8412–0906–5; paperback ISBN 0–8412–0933–2

Career Transitions for Chemists
By Dorothy P. Rodmann, Donald D. Bly, Frederick H. Owens, and Anne-Claire Anderson
240 pp; clothbound ISBN 0–8412–3052–8; paperback ISBN 0–8412–3038–2

Chemical Activities (student and teacher editions)
By Christie L. Borgford and Lee R. Summerlin
330 pp; spiralbound ISBN 0–8412–1417–4; teacher edition, ISBN 0–8412–1416–6

Chemical Demonstrations: A Sourcebook for Teachers, Volumes 1 and 2, Second Edition
Volume 1 by Lee R. Summerlin and James L. Ealy, Jr.
198 pp; spiralbound ISBN 0–8412–1481–6
Volume 2 by Lee R. Summerlin, Christie L. Borgford, and Julie B. Ealy
234 pp; spiralbound ISBN 0–8412–1535–9

The Internet: A Guide for Chemists
Edited by Steven M. Bachrach
360 pp; clothbound ISBN 0–8412–3223–7; paperback ISBN 0–8412–3224–5

Laboratory Waste Management: A Guidebook
ACS Task Force on Laboratory Waste Management
250 pp; clothbound ISBN 0–8412–2735–7; paperback ISBN 0–8412–2849–3

Reagent Chemicals, Eighth Edition
700 pp; clothbound ISBN 0–8412–2502–8

Good Laboratory Practice Standards: Applications for Field and Laboratory Studies
Edited by Willa Y. Garner, Maureen S. Barge, and James P. Ussary
571 pp; clothbound ISBN 0–8412–2192–8

For further information contact:
Order Department
Oxford University Press
2001 Evans Road
Cary, NC 27513
Phone: 1-800-445-9714 or 919-677-0977
Fax: 919-677-1303

Highlights from ACS Books

Desk Reference of Functional Polymers: Syntheses and Applications
Reza Arshady, Editor
832 pages, clothbound, ISBN 0–8412–3469–8

Chemical Engineering for Chemists
Richard G. Griskey
352 pages, clothbound, ISBN 0–8412–2215–0

Controlled Drug Delivery: Challenges and Strategies
Kinam Park, Editor
720 pages, clothbound, ISBN 0–8412–3470–1

Chemistry Today and Tomorrow: The Central, Useful, and Creative Science
Ronald Breslow
144 pages, paperbound, ISBN 0–8412–3460–4

Eilhard Mitscherlich: Prince of Prussian Chemistry
Hans-Werner Schutt
Co-published with the Chemical Heritage Foundation
256 pages, clothbound, ISBN 0–8412–3345–4

Chiral Separations: Applications and Technology
Satinder Ahuja, Editor
368 pages, clothbound, ISBN 0–8412–3407–8

Molecular Diversity and Combinatorial Chemistry: Libraries and Drug Discovery
Irwin M. Chaiken and Kim D. Janda, Editors
336 pages, clothbound, ISBN 0–8412–3450–7

A Lifetime of Synergy with Theory and Experiment
Andrew Streitwieser, Jr.
320 pages, clothbound, ISBN 0–8412–1836–6

Chemical Research Faculties, An International Directory
1,300 pages, clothbound, ISBN 0–8412–3301–2

For further information contact:
Order Department
Oxford University Press
2001 Evans Road
Cary, NC 27513
Phone: 1-800-445-9714 or 919-677-0977
Fax: 919-677-1303